Use R!

Series Editors

Robert Gentleman, 23andMe Inc., South San Francisco, USA

Kurt Hornik, Department of Finance, Accounting and Statistics, WU Wirtschaftsuniversität Wien, Vienna, Austria

Giovanni Parmigiani, Dana-Farber Cancer Institute, Boston, USA

This series of inexpensive and focused books on R is aimed at practitioners. Books can discuss the use of R in a particular subject area (e.g., epidemiology, econometrics, psychometrics) or as it relates to statistical topics (e.g., missing data, longitudinal data). In most cases, books combine LaTeX and R so that the code for figures and tables can be put on a website. Authors should assume a background as supplied by Dalgaard's Introductory Statistics with R or other introductory books so that each book does not repeat basic material.

How to Submit Your Proposal

Book proposals and manuscripts should be submitted to one of the publishing editors in your region per email – for the list of statistics editors by their location please see https://www.springer.com/gp/statistics/contact-us. All submissions should include a completed Book Proposal Form.

* * *

For general and technical questions regarding the series and the submission process please contact Faith Su (faith.su@springer.com) or Veronika Rosteck (veronika.rosteck@springer.com).

Constantino Antonio García Martínez •
Abraham Otero Quintana • Xosé A. Vila •
María José Lado Touriño •
Leandro Rodríguez-Liñares •
Jesús María Rodríguez Presedo •
Arturo José Méndez Penín

Heart Rate Variability Analysis with the R package RHRV

Second Edition

Constantino Antonio García Martínez
CEU San Pablo University
Madrid, Spain

Abraham Otero Quintana
CEU San Pablo University
Madrid, Spain

Xosé A. Vila
University of Vigo
Ourense, Spain

María José Lado Touriño
University of Vigo
Ourense, Spain

Leandro Rodríguez-Liñares
University of Vigo
Vigo, Spain

Jesús María Rodríguez Presedo
University of Santiago de Compostela
Santiago de Compostela, Spain

Arturo José Méndez Penín
University of Vigo
Ourense, Spain

ISSN 2197-5736   ISSN 2197-5744   (electronic)
Use R!
ISBN 978-3-031-65752-8   ISBN 978-3-031-65753-5   (eBook)
https://doi.org/10.1007/978-3-031-65753-5

Mathematics Subject Classification: 37M10, 65T50, 65T60, 62G10, 62F03, 65P20

© The Editor(s) (if applicable) and The Author(s), under exclusive license to Springer Nature Switzerland AG 2017, 2024

This work is subject to copyright. All rights are solely and exclusively licensed by the Publisher, whether the whole or part of the material is concerned, specifically the rights of translation, reprinting, reuse of illustrations, recitation, broadcasting, reproduction on microfilms or in any other physical way, and transmission or information storage and retrieval, electronic adaptation, computer software, or by similar or dissimilar methodology now known or hereafter developed.
The use of general descriptive names, registered names, trademarks, service marks, etc. in this publication does not imply, even in the absence of a specific statement, that such names are exempt from the relevant protective laws and regulations and therefore free for general use.
The publisher, the authors and the editors are safe to assume that the advice and information in this book are believed to be true and accurate at the date of publication. Neither the publisher nor the authors or the editors give a warranty, expressed or implied, with respect to the material contained herein or for any errors or omissions that may have been made. The publisher remains neutral with regard to jurisdictional claims in published maps and institutional affiliations.

This Springer imprint is published by the registered company Springer Nature Switzerland AG
The registered company address is: Gewerbestrasse 11, 6330 Cham, Switzerland

If disposing of this product, please recycle the paper.

Preface

The rhythm of the heart is influenced by both the sympathetic and parasympathetic branches of the Autonomic Nervous System. There are also some feedback mechanisms modulating the heart rate that try to maintain cardiovascular homeostasis by responding to the perturbations sensed by baroreceptors and chemoreceptors. Another major influence on the heart rate is the Respiratory Sinus Arrhythmia: the heartbeat synchronization with the respiratory rhythm. All these mechanisms are responsible for continuous variations in the heart rate of a healthy individual, even at rest. These variations are referred to as Heart Rate Variability (HRV). Subtle characteristics of these small variations conceal information about all the mechanisms underlying heart rate control, and hence about the health status of the individual.

Since the 1960s, researchers have developed a wide range of algorithms to extract the information hidden in these variations. Using these algorithms researchers have found markers for many pathologies such as myocardial infarction, diabetic neuropathy, sudden cardiac death, and ischemia. The starting point for all these algorithms is a simple recording of the instantaneous heart rate of the patient, usually extracted from an electrocardiogram. Therefore, a diagnostic based on an HRV marker is inexpensive, simple to perform, and requires no invasive procedure, making it a very attractive test. This is probably the reason behind the increasing amount of research related to HRV (see Fig. 1).

From the point of view of the authors, the main hindrance in the HRV research field is the difficulty in reproducing results from other researchers. When a new analysis technique or a new finding is published in the HRV literature, thinking it will be easy to reproduce the same result on your own data often is a mistake. We have tried it on several occasions. But the exact reproduction of the results was not possible, although we obtained results that qualitatively were similar to the originals. This is due to the lack of standardization in the values of many parameters and other implementation details in the HRV algorithms. Some examples are how exactly ectopic beats are filtered, the algorithm used to interpolate the RR intervals to obtain a time series of constant sampling frequency, how to remove the DC component (from all the RR series, from each window...), the window type

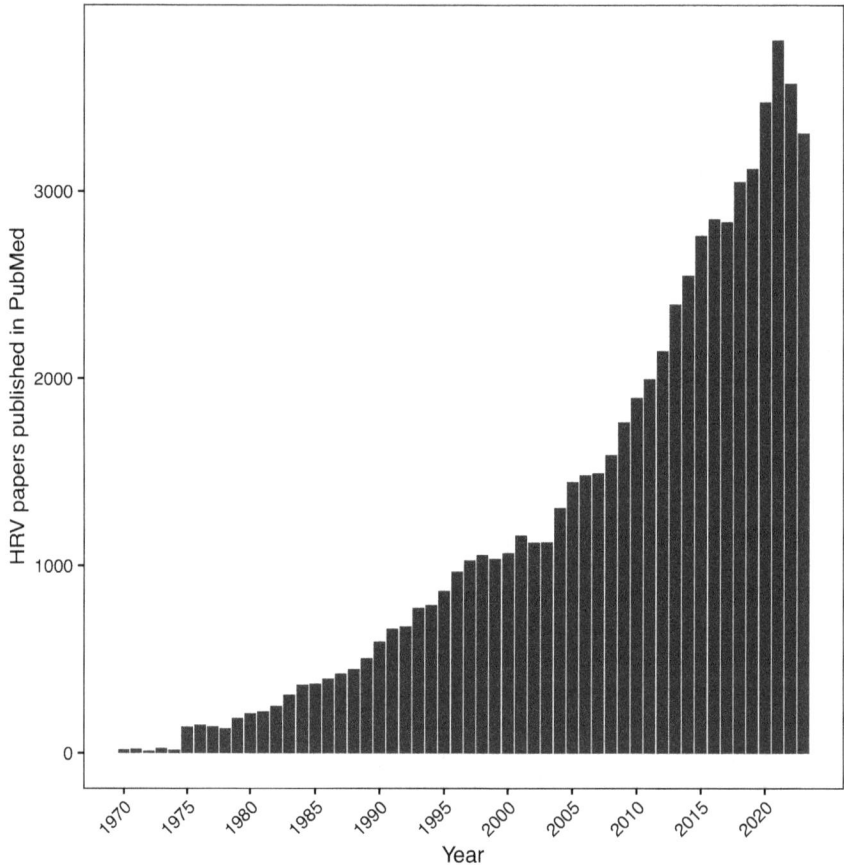

Fig. 1 Papers on HRV published per year according to PubMed

(Hamming, Welch,...), window size and window overlap used in the Fast Fourier Transform, or the mother wavelet used.

Any HRV analysis inherently involves dozens of decisions, both explicit and implicit, regarding parameters and implementation details. While documenting some of these choices can be challenging within a scientific paper, they are crucial for ensuring the faithful and accurate reproduction of the results. This challenge is further compounded when analyses rely on third-party tools whose source code is unavailable. In such cases, many decisions are made by the tool developers, potentially leaving researchers unaware of, or unable to control, some of these choices.

Another hindrance in the field is that researchers often use analytical techniques that are not the current state of the art, simply because their tool of choice does not support them, and they do not have the time and/or the necessary expertise to implement the techniques themselves. There is often a disconnect in the HRV literature between researchers who develop new and more powerful analysis techniques (often engineers), and those performing applied research in humans or animals (often physicians). The latter still often use older less powerful techniques and do not benefit from the progress made by the former. For example, in the literature there are many more HRV studies using the Fourier Transform than the Wavelet Transform, despite the theoretically superior properties of the latter for the analysis of nonstationary signals. We believe that the main reason for this is the historical lack of HRV analysis tools with support for spectral analysis based on wavelets.

RHRV is our attempt to address these problems. RHRV is a free of charge and open source package for the R environment that comprises a complete set of tools for Heart Rate Variability analysis. RHRV can import data files containing heartbeat positions in the most broadly used formats, and supports time domain, frequency domain, and nonlinear (fractal and chaotic) HRV analysis. The vast majority of the commonly used HRV analysis algorithms used in the literature have already been implemented in the tool. For example, the tool supports frequency analysis using the Fourier Transform (with and without Daniell smoothers), Short Time Fourier Transform, autoregressive models, Lomb-Scargle periodogram, and the Wavelet Transform. And we will continue adding new functionality to RHRV. Furthermore, as any good open source project, contributions are welcome.

Beyond being an invaluable help when performing HRV analysis (a typical HRV analysis with RHRV usually has just 10–15 lines of R code), we believe that RHRV can help the whole HRV field. Simply by publishing the RHRV analysis script as supplementary material of a paper the reproduction of the results over the same, or over new data, will be trivial: just run the script. Being RHRV an open and free package, no one should have any impediment to reproducing the results. And given that the state-of-the-art analysis techniques are implemented in RHRV, there is no reason not to use them. For example, in RHRV the difference between carrying out a spectral analysis based on Fourier or Wavelets is simply changing a parameter in a function call.

The main novelty of RHRV version 5.0, the version that corresponds to this book's second edition, is the addition of functionality for population-based HRV analysis. This functionality streamlines the analysis by automating not only the calculation of HRV-derived indices for multiple recordings but also the subsequent statistical analysis. The ultimate goal of HRV studies is often to identify statistically significant differences in HRV indices between populations. Traditionally, HRV analyses involve calculating indices for each recording within each population, followed by statistical tests to detect inter-population differences.

RHRV 5.0 introduces a new function, *RHRVEasy*. This function takes two or more folders as arguments, each containing RR recordings from a different population. *RHRVEasy* calculates temporal, frequency, and nonlinear domain indices for all recordings and performs a comprehensive statistical analysis to identify those presenting statistically significant differences between the populations. Remarkably, all of this can be done with a single line of R code. However, to achieve this simplification, some design decisions were made that might not be suitable for all analyses. Consequently, some researchers may prefer the greater control and flexibility offered by using the underlying RHRV functions directly, rather than opting for the simplicity and automation provided by *RHRVEasy*. This new functionality of RHRV is covered in the new Chap. 8 of this book.

As a minor but valuable addition to this book's second edition, all the code and data used are available at https://github.com/RHRV-team/RHRVBook/. We hope that the availability of these resources encourages a hands-on approach, enhancing the learning experience that this book aims to provide.

When the preface to the first edition of this book was written in early 2016 RHRV was downloaded about 450 times a month just from the RStudio's CRAN mirror. This number of downloads has continued to grow, and since 2020 the package has been downloaded on average more than 1,000 times per month (see Fig. 2). This

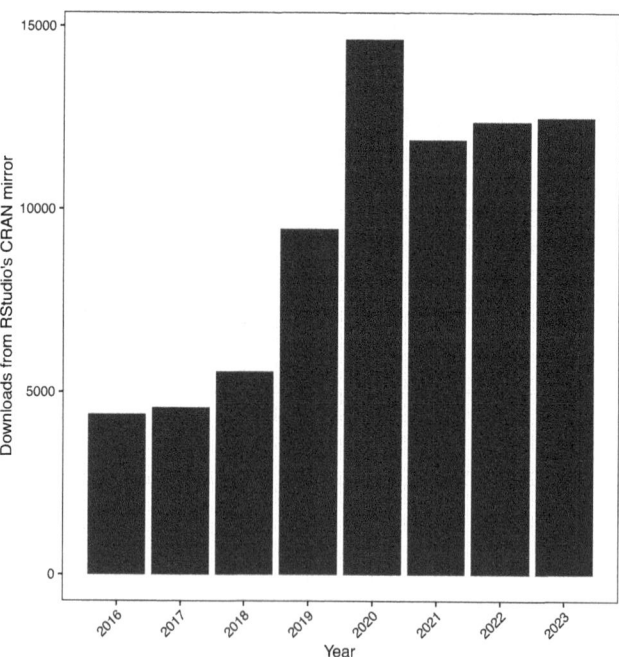

Fig. 2 RHRV downloads in the RStudio's CRAN mirror since 2016–2023

shows that many researchers have noticed the advantages of RHRV, and a strong community has already formed around it. We hope this trend continues, driven both by the second edition of this book and by the new functionality that RHRV 5 incorporates.

Madrid, Spain	Constantino Antonio García Martínez
Madrid, Spain	Abraham Otero Quintana
Ourense, Spain	Xosé A. Vila
Ourense, Spain	María José Lado Touriño
Vigo, Spain	Leandro Rodríguez-Liñares
Santiago de Compostela, Spain	Jesús María Rodríguez Presedo
Ourense, Spain	Arturo José Méndez Penín
May, 2024	

Contents

1 Introduction to Heart Rate Variability 1
  1.1 Historical Perspective 1
  1.2 Physiological Basis 4
      1.2.1 Cardiac Output and Heart Rate 4
      1.2.2 Autonomous Nervous System 4
      1.2.3 Autonomous Nervous System and Heart Rate Regulation ... 5
      1.2.4 Nonlinear Dynamics of the Heart 7
  1.3 Clinical Applications 7
      1.3.1 Monitoring 8
      1.3.2 Acute Care 10
      1.3.3 Chronic Disorders 11
  1.4 Software Tools for Heart Rate Variability Analysis 11
  References 15

2 Loading, Plotting, and Filtering RR Intervals 21
  2.1 Getting Started 21
  2.2 Data File Format 21
  2.3 Loading Beat Series into RHRV 23
  2.4 Preprocessing 26
      2.4.1 Instantaneous Heart Rate Signal Extraction 26
      2.4.2 Removing Artifacts 27
      2.4.3 Interpolation of the Heart Rate Signal 29
  2.5 Preprocessing Beat Data with RHRV 29
  References 33

3 Time-Domain Analysis 37
  3.1 Time-Domain Measures 37
  3.2 Time-Domain Analysis with RHRV 40
  3.3 Changes in HRV Time-Based Statistics Under Pathological
     Conditions 42
  References 43

4 Frequency-Domain Analysis ... 45
- 4.1 Frequency Components of the HRV ... 45
- 4.2 Frequency Analysis Techniques ... 47
  - 4.2.1 Frequency Analysis of Stationary Signals ... 47
  - 4.2.2 Frequency Analysis of Nonstationary Signals ... 49
- 4.3 Frequency-Domain Analysis with RHRV ... 51
  - 4.3.1 Frequency Analysis of Stationary Signals ... 51
  - 4.3.2 Frequency Analysis of Nonstationary Signals ... 62
- 4.4 Changes in HRV Frequency-Based Statistics Under Pathological Conditions ... 74
- References ... 75

5 Nonlinear and Fractal Analysis ... 79
- 5.1 An Overview of Nonlinear Dynamics ... 79
- 5.2 Chaotic Nonlinear Statistics ... 80
  - 5.2.1 Nonlinearity Tests ... 80
  - 5.2.2 Phase Space Reconstruction ... 81
  - 5.2.3 Correlation Dimension ... 82
  - 5.2.4 Generalized Correlation Dimension and Information Dimension ... 83
  - 5.2.5 Kolmogorov-Sinai Entropy ... 85
  - 5.2.6 Maximal Lyapunov Exponent ... 85
  - 5.2.7 Recurrence Quantification Analysis (RQA) ... 86
  - 5.2.8 Poincaré Plot ... 90
- 5.3 An Overview of Fractal Dynamics ... 90
  - 5.3.1 Detrended Fluctuation Analysis ... 92
  - 5.3.2 Power Spectral Density Analysis ... 93
- 5.4 Chaotic Nonlinear Analysis with RHRV ... 93
  - 5.4.1 Nonlinearity Tests ... 95
  - 5.4.2 Phase Space Reconstruction ... 98
  - 5.4.3 Nonlinear Statistics Computation ... 104
  - 5.4.4 Generalized Correlation Dimension and Information Dimension ... 106
  - 5.4.5 Sample Entropy ... 112
  - 5.4.6 Maximal Lyapunov Exponent ... 114
  - 5.4.7 RQA ... 116
  - 5.4.8 Poincaré Plot ... 119
- 5.5 Fractal Analysis with RHRV ... 120
  - 5.5.1 Detrended Fluctuation Analysis ... 121
  - 5.5.2 Power Spectral Analysis ... 123
- 5.6 Nonlinear and Fractal Analysis of HRV Under Pathological Conditions ... 124
- 5.7 Some Final Remarks Regarding HRV Analysis with Chaotic and Fractal Techniques ... 125
- References ... 127

6	**Comparing HRV Across Different Segments of a Recording**		131
	6.1	Episodes and Physiological Events	131
	6.2	Using Episodes in RHRV	132
		6.2.1 Managing Episodes in a HR Record	133
	6.3	Using Episodes in Plots	136
	6.4	Making Use of Episodes in HRV Analysis	138
	6.5	An Example	140
	6.6	Clinical Applications of HRV Analysis by Episodes	142
	References		144
7	**Putting It All Together: A Practical Example**		147
	7.1	Problem Statement	147
	7.2	Methodology	149
		7.2.1 Database Description	149
		7.2.2 Applying HRV Analysis	150
	References		157
8	**Automating HRV Analysis: *RHRVEasy***		159
	8.1	Introduction	159
	8.2	Time and Frequency Index Calculation with *RHRVEasy*	160
	8.3	Statistical Analysis	163
	8.4	Saving and Inspecting the Analysis Results	164
	8.5	A First Example	165
		8.5.1 Changing the Main Default Parameters	167
	8.6	Comparing More than Two Experimental Groups	169
	8.7	Nonlinear Index Calculation	172
	8.8	Parallelization of the Calculations	174
	8.9	Conclusions	175
	References		175
A	**Installing RHRV**		177
	A.1	RHRV Installation	177
	References		178
B	**How Do I Get a Series of RR Intervals from a Clinical/Biological Experiment?**		179
	B.1	QRS Detectors	179
	B.2	Signal Conditioning	181
	B.3	Creating a WFDB Compatible Record	182
	B.4	Creating a Beat Annotation File	183
	B.5	Extracting RR Intervals from an Annotation File	184
	B.6	Interpolation	185
	References		185
Index			187

Chapter 1
Introduction to Heart Rate Variability

Abstract Searching on Google Scholar the string "heart rate variability" (HRV) provides about half a million references, which gives us an idea of the research activity around this concept. This chapter describes the historical development of this research field, from the pioneering work of the eighteenth and nineteenth centuries to the boom of the final decade of the twentieth century. HRV analysis is important from a clinical point of view because of its direct relationship with the autonomic nervous system (ANS). Therefore, we include a section explaining the physiological basis of HRV and its relationship with the ANS. Finally, we review the most relevant clinical applications of HRV in three distinct lines: patient monitoring, acute care, and chronic disorders. The reader of this chapter must bear in mind that the purpose of this book is to present a software tool, RHRV, which greatly simplifies performing heart rate variability analysis. This chapter does not pretend to be a comprehensive review of the physiology of heart rate variation, but just a brief introduction to the field whose main purpose is building a common language to be able to present concepts more effectively throughout the book.

1.1 Historical Perspective

The heartbeat and pulse were already "known" by Greek physicians such as Galen of Pergamon (130–200 AD), who mentioned it on many of his manuscripts relating it to the "vital pneuma" and the "spirited soul." They knew that lack of pulse caused death and, based on its power, frequency, and regularity, they could make diagnoses. Since then, great progress has been made in the analysis of the variations of the heart rate (HR), which can be seen summarized in Fig. 1.1.

In the early eighteenth century, John Floyer (1649–1734) invented the "physician pulse watch" (1707), a precursor of the chronometer, which he began to use to accurately measure the heart rate of patients.

An important proof of the variability of the heart rate was contributed by Stephen Hales (1677–1761) who, in 1733, published a manuscript [1] in which he mentioned that the distance between heartbeats and the blood pressure varied during the respiratory cycle. Later, Carl Ludwig (1816–1895) improved these observations,

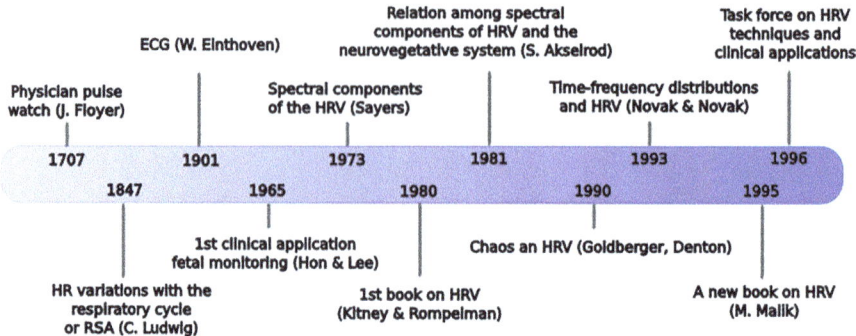

Fig. 1.1 Main heart rate variability (HRV) analysis historic milestones

clearly stating that the pulse increased during inspiration and decrease during expiration [2], thus laying the foundation of what would be called later respiratory sinus arrhythmia (RSA).

The transition from prehistory to modernity in the field of electrocardiography was given by Willem Einthoven (1860—1927), who recorded the first electrocardiogram (ECG) in 1901 using galvanometers [3]. The current nomenclature of (P, Q, R, S, T) was first used by Einthoven.

The ECG recording techniques improved gradually, and in 1961 Norman Holter (1914–1983) developed a device (that still bears his name, holter) allowing the registration of long-term ECG records [4].

From the 1970s on, improvements in measuring equipment and computational techniques caused this research field to take off. In 1970, Caldwell presented the "cardiotachometer" [5], capable of automatically recording beat positions using a small and inexpensive equipment. In 1973, Rompelman [6] presented another device capable of performing a basic statistical analysis of the HR values.

Due to the generalization of these devices and computers, specific studies on HRV started to appear, identifying the basic characteristics of the HR signal and isolating some well-defined peaks in its spectrum [6, 7]. In 1977, a paper comparing different techniques of HRV analysis was published [8], and in 1982, the first implementation of a system for HRV analysis over a standard PC was developed [9], which made this kind of analysis affordable for any researcher.

It could be considered that the first documented clinical application of HRV was the one published by Hon and Lee in 1963 [10]. They noted that fetal distress was accompanied by changes in HRV, even before there was a detectable change in HR. Another paper that boosted the research in this field was the one by Wolf in 1967 [11], which showed the relationship between HRV and the ANS and its role in sudden cardiac death.

It was certainly in the 1980s when the field of HRV reached maturity. Methodological issues were fixed, its physiological basis and its relationship with the ANS were demonstrated, and more clinical applications were described. In 1980, the first book devoted entirely to the HRV was published [12]. In 1981, Akselrod et al. published a paper in Science in which the relationship among each one of the spectral components of HRV and the various components of the ANS was empirically demonstrated [13]. A proof of the relevance of this paper is that it currently has more than 7600 citations in Google Scholar.

Other important paper was the one published by Pagani et al. in 1986 [14], describing results obtained with people and dogs of different age ranges following different therapies and placed in different positions. Although many other papers were published in this decade, we will cite only one more: the paper of Baselli et al. of 1987 [15], in which they describe the results of an analysis of hypertensive, diabetic, and myocardial infarction patients, compared with healthy people.

Until the 1990s, the parameters used for HRV analysis were limited to statistical parameters in the time domain and power in spectral bands obtained by the Fourier transform (FT). In 1990, two important works on fractality and chaotic dynamics of HR were published [16, 17], questioning the "linear" classical techniques and promoting the analysis of nonlinear dynamics of HR. Soon interesting clinical applications of these techniques were documented, such as in fetal monitoring [18]. Technological advances and applications of this line continue to the present.

It was also in this decade when some doubts about the validity of the Fourier transform as spectral estimator in some circumstances were raised. One of the first studies comparing classical Fourier estimation against nonparametric estimation and time-frequency distributions is the one by Pola et al. in 1996 [19]. Another important paper about this subject was published by Novak et al. in 1993 [20], which empirically established the superiority of time-frequency distributions when dealing with signals whose spectrum varies with time.

In the 1990s, another book devoted exclusively to HRV was published by Malik and Camm [21]. This book was a reference in the field of HRV for a long time and described as clinical applications of HRV the stratification of patients after myocardial infarction, its relationship with the occurrence of ventricular arrhythmias and sudden cardiac death, and its behavior in transplanted patients, in diabetic patients, and in fetal monitoring. Also in this decade, in 1996, the European Society of Cardiology and the North American Society of Pacing and Electrophysiology published a proposal for the methodology of HRV analysis prepared by a working group of experts [22]. Despite being an important contribution, it was quickly overtaken by technological advances since it did not consider using spectral estimators different from the FT, nor it included nonlinear analysis techniques [23].

Research in the field of HRV has not stopped in recent years. New methodologies have been proposed, experiments have been developed to better explain its physiological basis, and different studies have confirmed its clinical utility. It also has been applied to other fields such as psychiatry, sports training, or marketing. All of them will be discussed in the following sections.

1.2 Physiological Basis

1.2.1 Cardiac Output and Heart Rate

Cardiac output is defined as the volume of blood pumped by the heart in 1 minute. Therefore, this magnitude is in fact the product of two variables: the volume of blood pumped at each beat (which in turn is a function of blood pressure) and the heart rate.

Cardiac output varies greatly according to the physical constitution of the subjects. Obviously, it is not the same as pump blood to oxygenate and feed the cells of a person who is 1.90 m tall and weighs 85 kg, than to meet the needs of a person of 1.60 m and 58 kg [24]. Other factors, such as gender or age, also influence the cardiac output needs.

In addition to these factors, the cell needs of each individual change dynamically. Of course, necessities are not the same when sleeping or when running a marathon. However, without resorting to such extreme situations, simply sitting and getting up produces changes in these needs, and some adjustment in the cardiac output is required. The responsible for controlling cardiac output in this different settings is the autonomic nervous system (ANS) [25].

Even more, a subject in the same situation will have a cardiac output that is continuously changing, from second to second. In fact, as it was mentioned in the previous section, it is known since the nineteenth century that there is a variation in heart rate during each breathing cycle. In a healthy individual, the autonomic nervous system would properly make these continuous adjustments in heart rate and blood pressure, both the "fast" adjustments to meet "abrupt" situations and the long-term adjustments to maintain cardiac output in a stable situation. Since the mid-twentieth century, and as we have also pointed out in the previous section, we know that it is possible to use the heart rate variability as an indirect measure of the status of the autonomic nervous system.

Here we will briefly present the characteristics of the autonomic nervous system and then analyze its relationship with HRV.

1.2.2 Autonomous Nervous System

The autonomous or vegetative nervous system is responsible for regulating involuntary body responses (heart rate, digestion, pupil dilatation, etc.). It is also responsible for maintaining the body stable against external disturbances, which is known as homeostasis [26].

Within the autonomous nervous system, we distinguish three components:

- Sympathetic system: it uses acetylcholine and norepinephrine as neurotransmitters. It has nodes on both sides of the spinal column and in the aorta. It is involved

1.2 Physiological Basis

in situations that require energy expenditure and in stressful situations. It is responsible, for example, for increasing the heart rate in dangerous situations.
- Parasympathetic system: it uses acetylcholine as neurotransmitter and its nodes are mainly in the sacrum area and next to the brain. It is responsible for storing and preserving energy, to keep the body in a "normal" situation. Therefore, it would be also responsible for recovering the normal heart rate, when the danger has disappeared.
- Enteric nervous system: it is specifically responsible for controlling the gastrointestinal system.

As far as the heart rate is concerned, the parasympathetic system would be responsible for maintaining a stable heart rate value, enough to meet the required cardiac output, but without unnecessary waste. In fact, it seems proved that in the absence of autonomic control, "intrinsic" heart rate would be between 100 and 120 beats per minute [21]. By contrast, the sympathetic system would be responsible for raising the heart rate temporarily to prepare the body for situations that require increased energy expenditure. The greater or lesser activation of both systems and their different response times would be the cause of the continuous changes in heart rate, as we will discuss in the next section.

1.2.3 Autonomous Nervous System and Heart Rate Regulation

To explain this relationship, we will use the results of the experiment described by Akselrod et al. in Science in 1981 [13]. This paper was a key milestone in the use of HRV for clinical purposes.

In 1981, it was established that variations in heart rate were not just noise, and it was known that, if the series of "instantaneous heart rate" was taken as a time series, and its spectrum was analyzed, some predominant peaks in certain frequencies appeared. In their paper, Akselrod et al. quoted two pioneering works documenting this behavior. Authors mentioned the papers of Kitney-Rompelman and Sayers [7, 12], indicating that, besides the component related with the RSA (about 0.25 Hz), other low-frequency components could be distinguished, specifically around 0.04 Hz (low-frequency) and 0.12 Hz (mid-frequency). Moreover, they mentioned other papers by Hyndman et al. and Kitney [27, 28], suggesting that the 0.04 Hz frequency component could be related to cyclic fluctuations in peripherical vasomotor tone associated with termoregulation and that the 0.12 Hz component is related with the frequency response of the baroreceptor reflex.

Akselrod et al. designed an experiment to test whether there were indeed those three spectral components of HRV by using unanesthetized, conscious dogs, to which some drugs, whose effects on the ANS were known, were administered.

First, they administered glycopyrrolate to block the parasympathetic system observing that the mid- and high-frequency peaks were abolished, while the low-frequency peak was reduced. Then, they administered propranolol to block

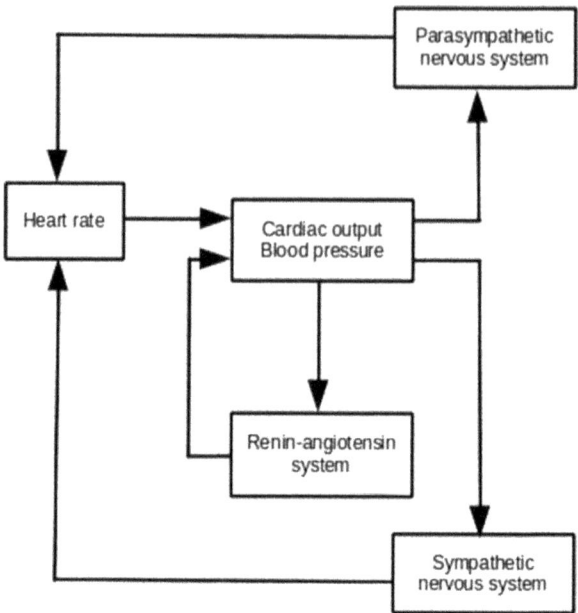

Fig. 1.2 The cardiovascular control model proposed in [13]

the sympathetic system and observed a reduction in the low-frequency peak. When administering both drugs, they noted the removal of all peaks, making the heart rate series steady, almost a metronome. Finally, investigators administered a nonapeptide-converting enzyme inhibitor to block the renin-angiotensin system and observed a large increase in the low-frequency peak.

To explain these results, they proposed, in the same paper, a simple model of "short-term cardiovascular control." In Fig. 1.2, we reproduce this model. According to their own explanation, the sympathetic and parasympathetic nervous systems were directly responsible for modulating heart rate in response to fluctuations in variables such as arterial blood pressure. The response time of the parasympathetic system is faster than the sympathetic one, so it is the only one that has effects on the peaks of medium and high frequency. However, the low-frequency peak is influenced by both systems. Finally, Akselrod et al. suggested that the renin-angiotensin system somehow limits fluctuations in the vasomotor tone, and when it is blocked, the vasomotor tone increases, occurring its manifestations mainly at low frequencies.

Akselrod's findings and their model are shared by many other authors, but some have added interesting information. For example, Saul [29] suggests that if the respiratory rate decreases, the high-frequency peak could scroll down, interacting with the medium-frequency peak and, therefore, entering the zone of influence of the sympathetic system. Malliani et al. [30] studied the occurrence of circadian rhythms when records of 24 hours were examined: they observed an increase in the low-frequency peak daytime, which intensifies with physical exercise.

1.2.4 Nonlinear Dynamics of the Heart

In the previous section, some heart rate spectral components were mentioned. To obtain these components, we start from the premise that analyzed data correspond to a stationary and linear system, which allows us to use techniques such as the FT to obtain its spectrum.

For decades that assumption has been questioned. In fact, it has been found by means of mathematical models and by developing experiments with real heart data that a healthy heart has a nonlinear, nonperiodic behavior, showing long-range correlations and typical fractal system features [17, 31]. Moreover, it is known that this nonlinear behavior is a symptom of cardiovascular health and that certain diseases or conditions, such as old age, diminish such nonlinearity, which is considered an indicator of worsening prognosis.

In addition, it is believed that when the cardiovascular system becomes more linear, and therefore more predictable, its ability to adapt to changes in the environment decreases, thus increasing the risk of falling into situations such as arrhythmias or fibrillations [17].

1.3 Clinical Applications

Given that there is a relationship between HRV and the control of the cardiac activity by the neurovegetative control system, it seems logical that in any illness in which the ANS is involved, HRV will be affected in some way. Finding associations between HRV alterations and pathologies is hindered by the fact that there are many causes that may affect HRV, even in healthy people. For example, the body position while the ECG recording is done, if it is done at daytime or nighttime, and the mood of the person. This is the reason why it is so difficult to conclude the existence of any disease or aggravation by studying only the value of a single HRV index.

The most relevant work done so far in order to standardize methodologies for the analysis of HRV and to establish clinical protocols using that information was published in 1996 by the European Society of Cardiology and the North American Society of Pacing and Electrophysiology [22]. In this report, although several potential clinical applications of HRV are cited, only two with a broad consensus are mentioned: predicting the risk after myocardial infarction and its use as an early detector of diabetic neuropathy. A recent publication by Sassi et al. [23] reviewed new appearing methodologies to analyze HRV, but none of them has significantly contributed to the development of new clinical tools.

A later article about clinical applications of HRV [32] added a third potential application: its use in risk stratification after a heart failure.

In the important book published by Malik and Camm in 1995 [21], the following clinical applications are also mentioned:

- Prediction of ventricular arrhythmias
- Prediction of sudden death
- Hypertension monitoring
- Prognosis of patients with heart transplant
- Assessment of fetal status
- Assessment of diabetic patients

However, in none of these cases, neither clear clinical guidelines are given nor protocols with broad consensus are established. This prevents HRV from being used in routine clinical practice as a marker for these conditions.

One of the more recent books about the clinical utility of HRV is the one by Kamath et al. [33]. In this book, three possible scenarios in which the HRV could be used to obtain relevant information from a clinical point of view are presented. These scenarios are patient monitoring, application in intensive care units (ICUs), and use in some chronic diseases.

In the rest of this section, based on the chapters of this book and other recent papers, we provide a detailed explanation about some of these clinical applications.

1.3.1 Monitoring

1.3.1.1 Fetal Monitoring

The fetus heartbeat can be distinguished in a simple abdominal ECG, superimposed to the mother ECG. There are methods to separate both signals [33]. A simple visual representation of the fetal heart rate shows two fundamental differences with the adult. First, its average value is higher, around 150 bpm (beats per minute). Second, periodical accelerations and decelerations every few minutes are superimposed over the short-term variability [21]. It is also known that as the fetus matures, the heart rate decreases and its complexity or variability increases.

A decrease in the fetal HRV is considered negative, and correlations have been found between its descent and fetal metabolic acidosis, neurological abnormalities, drug effects, or growth delays [21].

In a recent paper by Nageotte [34], fetuses were classified as follows, depending on their HR and HRV:

- Type I or normal, with HR between 110 and 160 bpm and moderate HRV
- Type III or abnormal, with HR outside those ranges and low HRV
- Type II or unclassified, those that do not fall into the other two categories

This classification has been recognized by various associations and is reported in various clinical guidelines. Its main drawback is that about 80% of fetuses were classified as Type II. More scientific studies and clearer clinical guides are needed in order to increase the clinical interest of fetal HRV.

1.3.1.2 HRV and Hypertension

Hypertension is a well-known factor of cardiac risk and its prevalence is high in Western society. It is known that hypertensive patients show alterations in their cardiovascular control system. Specifically, in these patients, the balance between sympathetic and parasympathetic systems is shifted toward the first. Since HRV is controlled by the autonomic nervous system, variations are expected when comparing the HRV of hypertensive patients with the healthy population. It also seems possible that HRV indices could be used in hypertensive patients' monitoring, as indicators of its evolution and prognosis, or even as an early detector of hypertension risk.

Studies developed until now indicate that there is a clear correlation between diminished HRV and risk of hypertension. It has also been found that HRV indices are reduced in hypertensive patients when compared to the healthy population [33]. Other studies [35] distinguish between patients with moderate and severe hypertension and found significant differences, enabling then correlation of HRV index with the severity of hypertension.

1.3.1.3 Sleep Disorders

During sleep, a healthy person goes through different stages, each one characterized by a different brain activity pattern. This has implications over the cardiac activity, which we know is mediated by the nervous system. Several studies have found correlations between sleep stages and changes in HRV indices [33, 36]. Another factor that causes alterations on HRV indices is the circadian rhythm itself [37], with differences between daytime and nighttime values.

It is therefore not surprising that some studies try to analyze the extent to which healthy people with sleep disorders show altered HRV patterns. Several of the authors of this book have participated in studies related to sleep apnea [38] and COPD (chronic obstructive pulmonary disease) [39], finding significant differences in some HRV indices. For example, when studying sleep apnea, we found differences not only between healthy people and patients with apnea but also among patients with different degrees of apnea.

1.3.2 Acute Care

1.3.2.1 HRV in the Intensive Care Unit (ICU)

Patients admitted to ICUs are permanently monitored. Variables such as temperature, ECG, oxygen saturation, respiration, and urine output are continuously recorded, and any deviation from the normal range usually triggers the corresponding alarm. Besides analyzing these variables, it is critical to diagnose the patient and to continually adjust his/her medication.

Many of these variables are controlled by the autonomic nervous system. Therefore, a malfunction of this system may cause alterations in them. In fact, it is known that such dysfunction occurs in patients with brain damage, infection, cardiovascular disorders, and multiorgan failure [33]. Therefore, an indirect measure of the state of the autonomic nervous system could be useful when evaluating or establishing the prognosis of such patients.

For example, in patients hospitalized for trauma, a low HRV correlates with increased mortality [40]. There are also studies that find low HRV in patients with sepsis who then evolve negatively [41].

1.3.2.2 HRV and Anesthesia

The application of anesthesia, whether local or general, is always a situation of increased risk, because it involves external interference in the control mechanism of the body's physiological variables. Therefore, especially in the application of general anesthesia, the patient is continuously monitored, and the anesthesiologist continuously monitors these variables to set the optimal dose of anesthesia.

Among the variables that the anesthesiologist monitors are heart rate, oxygen saturation, blood pressure, and respiratory rate. In the last decades, there have been experiments in which HRV was analyzed during the application of anesthesia [33], and significant variations in different HRV indices have been found.

For example, it has been found that a previous study of HRV can detect people at risk of hypotension when anesthetized for cesarean [42]. The effects of different drugs on HRV indices are also known, and studies exist on risk stratification after anesthesia. However, studies trying to correlate HRV indices and the ideal dose of anesthesia are not conclusive [43].

1.3.3 Chronic Disorders

1.3.3.1 Congestive Heart Failure

It is fully accepted that HRV decreases in patients with heart failure. These patients are less able to regulate their HR than the normal population, which is reflected in a lower variability [44].

In fact, decreased HRV has been used in many studies as a predictor of risk of heart failure in patients after myocardial infarction [45]. Going a step further, correlations between levels of severity of heart failure and HRV indices have been found; they have enabled the proposal of remote diagnostics systems to classify these patients [46]. The effects of various treatments on HRV indices have even been studied, including implantable pacemaker installation, supply of drugs, and prescription of moderate physical activity [33]. In all these cases, there has been recovery in HRV indices in parallel with the improvement of the cardiovascular system of the patients.

1.3.3.2 Neuropsychiatric Disorders

In today's society, psychiatric disorders have a growing prevalence. Obviously, all these disorders are associated with a malfunction of the nervous system and a loss of its capacity to adapt to the environment, due to a decrease in its flexibility.

HRV is one of the external manifestations of the nervous control system. In a healthy person, HRV is high and varies according to changes in environmental situations. But in a person with psychiatric disorders, the HRV is usually diminished, in accordance with his decrease.

Among the diseases for which there are studies that correlate HRV and disease severity are schizophrenia, bipolar disorder, anxiety disorders, post-traumatic stress disorders, depression, or suicidal ideation [33, 47–49].

1.4 Software Tools for Heart Rate Variability Analysis

Until recently, HRV was not considered essential by medical device manufacturers. As a consequence, its analysis was not included in their equipment. In the 1990s, existing monitoring systems basically included RR interval measures after QRS detection and a limited quantification of the HRV [50]. Others were based on microcomputers and allowed measures of a reduced number of HRV-related parameters (mean heart rate, mean rate deviation, and standard deviation of the RR intervals) [51]. However, in recent decades, as HRV has becoming increasingly important to detect and diagnose varied medical conditions, different software tools to automatically perform HRV analyses have arisen in both commercial and research fields. Two different general categories of tools can be mentioned. On the one hand, many apps for mobile devices and smartwatches are continuously appearing, mostly centered on non-healthcare personnel nor researchers, which only desire to control their wellness or fitness. On the other hand, desktop applications are very often used by clinicians and investigators, aiming at improving knowledge about HRV, as a further step in the diagnosis and detection of diseases and recovery processes.

Related to the applications for smartwatches and mobile devices, both for Android and iOS, most of them calculate the rMSSD parameter employing photoplethysmography (PPG) to record heart rate data, such as the Apple Watch [52], or the Fitbit watches series [53], which offer similar functionalities. The commercial Beats Analysis app for wearable Garmin devices [54] also calculates the rMSSD parameter, as well as the Elite-HRV [55] and the Welltory app, which also provide information about different time and frequency HRV parameters, such as rMSSD, SDNN, mean HR, pNN50, LF, or HF, among others [56] (all these parameters will be explained in detail in Chap. 3). Another app is VARSE, which allows to acquire HR signals from different Bluetooth devices and performs both real-time and frequency HRV analyses. VARSE is open source and can be installed via Google Play [57].

Limitations of most of these tools to analyze HRV are significant. The majority of former applications are designed for recreational purposes, but they do not serve for medical or investigation tasks, because heart data acquisition may not be sufficiently precise, or they can lack information in the HRV analyses [58, 59]. However, it has been demonstrated that some of them, such as Elite HRV or Welltory, provide similar values for short- and ultrashort HRV measures to those obtained when employing the ECG gold standard [60, 61]. The VARSE app has also been tested with other software packages, comparing HR data acquisition and HRV analysis results [62].

In relation to desktop applications, we have focused on the applications that are being used by researchers, and therefore have bibliographic citations, and on the commercial applications most used in the clinical setting. Table 1.1 shows their main characteristics.

Among the open-source tools, the PhysioNet Cardiovascular Signal Toolbox (PCST) processes data from RR intervals, as well as raw waveform data, and allows nonlinear analysis (all tools permits time and frequency analysis) [63]. Another interesting feature is that it allows automatic processing of batches of records.

RHRV, the tool covered in this book, is also a popular HRV software, developed in R language, that performs linear, nonlinear and time-frequency HRV analysis [64]. It can be freely downloaded [65]. It also permits batch processing and another interesting feature: the possibility of defining intervals of interest in the record (episodes) and comparing HRV indices inside and outside these intervals.

Many desktop applications to analyze HRV are provided with GUI. In these cases, non-experienced programming users can profit from the benefits of the software in an easy and simple way. However, some of the tools present a major drawback, since they are implemented in proprietary languages, such as Matlab. Thus, researchers need access to this commercial software if they want to add or implement new algorithms. An example can be HRVAS [66]. It includes nonlinear

1.4 Software Tools for Heart Rate Variability Analysis

Table 1.1 Software tools for HRV analysis

Name	Open source	Input	Nonlinear	Time-freq	GUI	Batch	Episodes	Update	Cites
PCST	Y(Matlab)	RR,ECG	Y	N	N	Y	N	2019	233
RHRV	Y	RR	Y	Y	N	Y	Y	2024	129
gHRV	Y	RR	Y	Y	Y	N	Y	2020	71
HRVAS	Y(Matlab)	RR	Y	Y	Y	Y	N	2018	166
RR-APET	Y	RR,ECG	Y	N	Y	Y	N	2020	17
ARTiiFACT	N(free)	RR,ECG	N	N	Y	N	N	2021	336
Kardia	Y(Matlab)	RR	Y	N	Y	N	Y	2022	124
Sinus Cor	Y(Matlab)	RR,ECG	N	Y	Y	N	N	–	47
Kubios Lite	N(free)	RR	Y	N	Y	N	N	2023	2688
Kubios	N	RR,ECG	Y	Y	Y	N	N	2023	2688
aHRV	N	RR,ECG	N	Y	Y	N	N		–
LabView	N	RR,ECG	N	N	Y	N	N		–
AcqKnowledge	N	ECG	N	N	Y	N	N		–

and time-frequency analysis and the option to do batch processing. It seems like an outdated project because the last update is from 2018.

ARTiiFACT was also developed in Matlab, but no Matlab license is required for executing the program [67]. It accepts RR and ECG records and can be freely downloaded [68].

Another open-source (but Matlab-dependent) tool is Kardia [69]. It provides functions for estimation of time, frequency, and nonlinear HRV parameters. It also offers the possibility of analyzing episodes within the record [70].

SinusCor is also another Matlab-based software, which performs time-frequency analyses [71]. It is also distributed under freeware license [72].

Related to free, open-source GUI applications, not Matlab dependent, we mention two, both of them developed in Python language. First is the gHRV software package [73]. This software implements nonlinear and time-frequency analyses and can be freely downloaded [74]. Second is RR-APET, a multiplatform tool that includes nonlinear analysis and permits batch processing [75]. Software is also available from GitLab [76].

The application most used by HRV researchers is undoubtedly Kubios [77], with more than 2000 citations in Google Scholar. Although it started out as a free application, it is currently distributed with a paid license. Its characteristics can be seen in Table 1.1: it does not allow the processing of several records in batch mode, nor does it contemplate the comparative study of episodes within the record.

They do, however, offer a free version called Kubios Lite [78], with limited functionality and for noncommercial use. It only accepts RR records and does not allow time-frequency analysis, among other limitations.

Other commercial tool, widely used and designed to perform HRV analysis, is aHRV [79]. This software for Windows, developed by Nevrokard and provided with a GUI, implements time, frequency, and time-frequency analysis.

The LabVIEW Biomedical Toolkit API [80], developed by National Instruments, also performs time and frequency analysis employing the HRV analysis VIs (virtual instruments) in a simple and easy way, even for non-experienced users, through a friendly GUI. Another proprietary software for HRV analysis is AcqKnowledge®, the one developed by BIOPAC Systems, Inc. [81].

Comparing the RHRV application with those listed in the Table 1.1, it should be noted that it is the only one that allows analysis in the time, frequency, time-frequency, and nonlinear domains and at the same time has sufficient flexibility to allow the analysis of batches of records and the comparative study of different episodes within the same record.

References

1. S. Hales, *Statistical Essays: Concerning Haemastaticks* (W. Innys and R. Manby, London, 1733). [Online]. Available: https://doi.org/10.5962/bhl.title.106596
2. C. Ludwig, Beiträge zur kenntniss des einflusses der respirationsbewegungen auf den blutlauf im aortensysteme. Arch. für Anat. Physiol. Wissenschaftliche Medicin **13**, 242–302 (1847)
3. W. Einthoven, Über die form des menschlichen electrocardiogramms. Pflügers Archiv Eur. J. Physiol. **60**(3), 101–123 (1895). [Online]. Available: https://doi.org/10.1007/BF01662582
4. N.J. Holter, New method for heart studies continuous electrocardiography of active subjects over long periods is now practical. Science **134**(3486), 1214–1220 (1961). [Online]. Available: https://doi.org/10.1126/science.134.3486.1214
5. W.M. Caldwell, L.D. Smith, M.F. Wilson, A wide-range linear beat-by-beat cardiotachometer. Med. Biol. Eng. **8**(2), 181–185 (1970). [Online]. Available: https://doi.org/10.1007/bf02509328
6. O. Rompelman, W. Snoeijer, H. Ros, A special purpose computer for dynamic statistical analysis of RR intervals, in *Digest of the 10th International Conference on Medical & Biological Engineering, Dresden*, vol. 31 (1973). [Online]. Available: https://doi.org/10.1007/BF02441043
7. B. Sayers, Analysis of heart rate variability. Ergonomics **16**(1), 17–32 (1973). [Online]. Available: https://doi.org/10.2174/1573403x16999201231203854
8. O. Rompelman, A. Coenen, R. Kitney, Measurement of heart-rate variability: part 1: comparative study of heart-rate variability analysis methods. Med. Biol. Eng. Comput. **15**(3), 233–239 (1977). [Online]. Available: http://doi.org/10.1007/BF02441043
9. O. Rompelman, J.B. Snijders, C.J. Van Spronsen, The measurement of heart rate variability spectra with the help of a personal computer. IEEE Trans. Biomed. Eng. **BME-29**(7), 503–510 (1982). [Online]. Available: https://doi.org/10.1109/tbme.1982.324922
10. E.H. Hon, S. Lee, Electronic evaluation of the fetal heart rate. VIII. Patterns preceding fetal death, further observations. Amer. J. Obstet. Gynecol. **87**, 814–826 (1963)
11. S. Wolf, The end of the rope: the role of the brain in cardiac death. Can. Med. Assoc. J. **97**(17), 1022 (1967)
12. R. Kitney, O. Rompelman, *The Study of Heart-Rate Variability*, no. 3 (Oxford, Oxford University Press, 1980). [Online]. Available: https://doi.org/10.1113/expphysiol.1981.sp002546
13. S. Akselrod, D. Gordon, F.A. Ubel, D.C. Shannon, A. Berger, R.J. Cohen, Power spectrum analysis of heart rate fluctuation: a quantitative probe of beat-to-beat cardiovascular control. Science **213**(4504), 220–222 (1981)
14. M. Pagani, F. Lombardi, S. Guzzetti, O. Rimoldi, R. Furlan, P. Pizzinelli, G. Sandrone, G. Malfatto, S. Dell'Orto, E. Piccaluga, Power spectral analysis of heart rate and arterial pressure variabilities as a marker of sympatho-vagal interaction in man and conscious dog. Circul. Res. **59**(2), 178–193 (1986). [Online]. Available: https://doi.org/10.1161/01.RES.59.2.178
15. G. Baselli, S. Cerutti, S. Civardi, F. Lombardi, A. Malliani, M. Merri, M. Pagani, G. Rizzo, Heart rate variability signal processing: a quantitative approach as an aid to diagnosis in cardiovascular pathologies. Int. J. Bio-Med. Comput. **20**(1), 51–70 (1987). [Online]. Available: https://doi.org/10.1016/0020-7101(87)90014-6
16. T.A. Denton, G.A. Diamond, R.H. Helfant, S. Khan, H. Karagueuzian, Fascinating rhythm: a primer on chaos theory and its application to cardiology. Amer. Heart J. **120**(6), 1419–1440 (1990). [Online]. Available: https://doi.org/10.1016/0002-8703(90)90258-Y
17. A.L. Goldberger, D.R. Rigney, B.J. West, Chaos and fractals in human physiology. Sci. Am. **262**(2), 42–49 (1990). [Online]. Available: https://doi.org/10.1038/scientificamerican0290-42
18. S.M. Pincus, R.R. Viscarello, Approximate entropy: a regularity measure for fetal heart rate analysis. Obstet. Gynecol. **79**(2), 249–255 (1992)
19. S. Pola, A. Macerata, M. Emdin, C. Marchesi, Estimation of the power spectral density in non-stationary cardiovascular time series: assessing the role of the time-frequency representations

(TFR). IEEE Trans. Biomed. Eng. **43**(1), 46–59 (1996). [Online]. Available: https://doi.org/10.1109/10.477700
20. P. Novak, V. Novak, Time/frequency mapping of the heart rate, blood pressure and respiratory signals. Med. Biol. Eng. Comput. **31**(2), 103–110 (1993). [Online]. Available: https://doi.org/10.1007/bf02446667
21. M. Malik, A.J. Camm, *Heart Rate Variability* (Futura Publishing Company, Austin, 1995). [Online]. Available: https://doi.org/10.1002/clc.4960130811
22. Task Force of the European Society of Cardiology and the North American Society of Pacing and Electrophysiology, Heart rate variability: standards of measurement, physiological interpretation and clinical use. Eur. Heart J. **17**, 354–381 (1996). [Online]. Available: https://doi.org/10.1161/01.CIR.93.5.1043
23. R. Sassi, S. Cerutti, F. Lombardi, M. Malik, H.V. Huikuri, C.-K. Peng, G. Schmidt, Y. Yamamoto, B. Gorenek, G.H. Lip, G. Grassi, G. Kudaiberdieva, J.P. Fisher, M. Zabel, R. Macfadyen, Advances in heart rate variability signal analysis: joint position statement by the e-Cardiology ESC Working Group and the European Heart Rhythm Association co-endorsed by the Asia Pacific Heart Rhythm Society. Europace **17**, euv015 (2015). [Online]. Available: https://doi.org/10.1093/europace/euv015
24. T. Collis, R.B. Devereux, M.J. Roman, G. de Simone, J.-L. Yeh, B.V. Howard, R.R. Fabsitz, T.K. Welty, Relations of stroke volume and cardiac output to body composition: the strong heart study. Circulation **103**(6), 820–825 (2001). [Online]. Available: https://doi.org/10.1161/01.cir.103.6.820
25. G.D. Thomas, Neural control of the circulation. Adv. Physiol. Edu. **35**(1), 28–32 (2011). [Online]. Available: https://doi.org/10.1152/advan.00114.2010
26. W. Janig, Autonomic nervous system, in *Human Physiology*, ed. by R.F. Schmidt, G. Thews (Springer, Berlin, 1989), pp. 333–370. [Online]. Available: https://doi.org/10.1007/978-3-642-73831-9_16
27. B. Hyndman, R. Kitney, B.M. Sayers, Spontaneous rhythms in physiological control systems. Nature **233**, 339–341 (1971). [Online]. Available: https://doi.org/10.1038/233339a0
28. R. Kitney, An analysis of the nonlinear behaviour of the human thermal vasomotor control system. J. Theor. Biol. **52**(1), 231–248 (1975). [Online]. Available: https://doi.org/10.1016/0022-5193(75)90054-5
29. J.P. Saul, Beat-to-beat variations of heart rate reflect modulation of cardiac autonomic outflow. Physiology **5**(1), 32–37 (1990). [Online]. Available: https://doi.org/10.1152/physiologyonline.1990.5.1.32
30. A. Malliani, M. Pagani, F. Lombardi, S. Cerutti, Cardiovascular neural regulation explored in the frequency domain. Circulation **84**(2), 482–492 (1991). [Online]. Available: https://doi.org/10.1161/01.cir.84.2.482
31. A.L. Goldberger, L.A. Amaral, J.M. Hausdorff, P.C. Ivanov, C.-K. Peng, H.E. Stanley, Fractal dynamics in physiology: alterations with disease and aging. Proc. Natl. Acad. Sci. **99**(Suppl 1), 2466–2472 (2002). [Online]. Available: https://doi.org/10.1073/pnas.012579499
32. A. Stys, T. Stys, Current clinical applications of heart rate variability. Clin. Cardiol. **21**(10), 719–724 (1998). [Online]. Available: https://doi.org/10.1002/clc.4960211005
33. M.V. Kamath, M. Watanabe, A. Upton, *Heart Rate Variability (HRV) Signal Analysis: Clinical Applications* (CRC Press, Boca Raton, 2012). [Online]. Available: https://doi.org/10.1201/b12756
34. M.P. Nageotte, Fetal heart rate monitoring. Semin. Fetal Neonatal Med. **20**(3), 144–148 (2015). [Online]. Available: https://doi.org/10.1016/j.siny.2015.02.002
35. H. Mussalo, E. Vanninen, R. Ikäheimo, T. Laitinen, M. Laakso, E. Länsimies, J. Hartikainen, Heart rate variability and its determinants in patients with severe or mild essential hypertension. Clin. Physiol. **21**(5), 594–604 (2001). [Online]. Available: https://doi.org/10.1046/j.1365-2281.2001.00359.x

References

36. I. Berlad, A. Shlitner, S. Ben-Haim, P. Lavie, Power spectrum analysis and heart rate variability in stage 4 and REM sleep: evidence for state-specific changes in autonomic dominance. J. Sleep Res. **2**(2), 88–90 (1993). [Online]. Available: https://doi.org/10.1111/j.1365-2869.1993.tb00067.x
37. H.V. Huikuri, M. Niemelä, S. Ojala, A. Rantala, M. Ikäheimo, K. Airaksinen, Circadian rhythms of frequency domain measures of heart rate variability in healthy subjects and patients with coronary artery disease. Effects of arousal and upright posture. Circulation **90**(1), 121–126 (1994). [Online]. Available: https://doi.org/10.1161/01.cir.90.1.121
38. M.J. Lado, A.J. Méndez, L. Rodríguez-Liñares, A. Otero, X.A. Vila, Nocturnal evolution of heart rate variability indices in sleep apnea. Comput. Biol. Med. **42**(12), 1179–1185 (2012). [Online]. Available: https://doi.org/10.1016/j.compbiomed.2012.09.009
39. C. Zamarrón, M.J. Lado, T. Teijeiro, E. Morete, X.A. Vila, P.F. Lamas, Heart rate variability in patients with severe chronic obstructive pulmonary disease in a home care program. Technol. Health Care **22**(1), 91–98 (2014). [Online]. Available: https://doi.org/10.3233/thc-140777
40. M.L. Ryan, M.P. Ogilvie, B.M. Pereira, J.C. Gomez-Rodriguez, R.J. Manning, P.A. Vargas, R.C. Duncan, K.G. Proctor, Heart rate variability is an independent predictor of morbidity and mortality in hemodynamically stable trauma patients. J. Trauma Acute Care Surgery **70**(6), 1371–1380 (2011). [Online]. Available: https://doi.org/10.1097/ta.0b013e31821858e6
41. D. Barnaby, K. Ferrick, D.T. Kaplan, S. Shah, P. Bijur, E.J. Gallagher, Heart rate variability in emergency department patients with sepsis. Acad. Emerg. Med. **9**(7), 661–670 (2002). [Online]. Available: https://doi.org/10.1111/j.1553-2712.2002.tb02143.x
42. R. Hanss, B. Bein, T. Ledowski, M. Lehmkuhl, H. Ohnesorge, W. Scherkl, M. Steinfath, J. Scholz, P.H. Tonner, Heart rate variability predicts severe hypotension after spinal anesthesia for elective cesarean delivery. Anesthesiol.-Hagerstown **102**(6), 1086–1093 (2005). [Online]. Available: https://doi.org/10.1097/00000542-200506000-00005
43. A. Mazzeo, E. La Monaca, R. Di Leo, G. Vita, L. Santamaria, Heart rate variability: a diagnostic and prognostic tool in anesthesia and intensive care. Acta Anaesthesiol. Scandinavica **55**(7), 797–811 (2011). [Online]. Available: https://doi.org/10.1111/j.1399-6576.2011.02466.x
44. G. Casolo, Heart rate variability in patients with heart failure, in *Heart Rate Variability* (Futura Publishing Company, Austin, 1995), pp. 499–46633
45. H.V. Huikuri, P.K. Stein, Clinical application of heart rate variability after acute myocardial infarction. Front. Physiol. **3**, (2012). [Online]. Available: https://doi.org/10.3389/fphys.2012.00041
46. L. Pecchia, P. Melillo, M. Bracale, Remote health monitoring of heart failure with data mining via CART method on HRV features. IEEE Trans. Biomed. Eng. **58**(3), 800–804 (2011). [Online]. Available: https://doi.org/10.1109/tbme.2010.2092776
47. K. Latalova, J. Prasko, T. Diveky, D. Kamaradova, A. Grambal, D. Jelenova, B. Mainerova, M. Cerna, M. Ociskova, H. Velartova et al., Euthymic bipolar affective disorder patients and their heart rate variability. Eur. Psych. **28**, 1 (2013). [Online]. Available: http://dx.doi.org/10.1016/S0924-9338(13)76016-4
48. A.J. Méndez, M.J. Lado, X.A. Vila, L. Rodríguez-Liñares, R.A. Alonso, A. García-Caballero, Heart of darkness: Heart rate variability on patients with risk of suicide, in *8th Iberian Conference on Information Systems and Technologies (CISTI)* (2013), pp. 1–4
49. G. Tan, T.K. Dao, L. Farmer, R.J. Sutherland, R. Gevirtz, Heart rate variability (HRV) and posttraumatic stress disorder (PTSD): a pilot study. Appl. Psychophysiol. Biofeedback **36**(1), 27–35 (2011). [Online]. Available: https://doi.org/10.1007/s10484-010-9141-y
50. F. Pinciroli, R. Tresca, A microprocessor-based device for synthesis of RR interval values during ambulatory monitoring. J. Med. Eng. Technol. **7**(5), 247–251 (1983). [Online]. Available: https://doi.org/10.3109/03091908309032593
51. S. Cahill, D.G. McClure, A microcomputer-based heart-rate variability monitor. IEEE Trans. Biomed. Eng. **BME-30**(2), 87–93 (1983). [Online]. Available: https://doi.org/10.1109/tbme.1983.325202

52. Apple Watch. Last Accessed 26 Feb 2024. [Online]. Available: https://www.apple.com/uk/watch/
53. Fitbit Watch. Last Accessed 26 Feb 2024. [Online]. Available: https://www.fitbit.com/global/us/home
54. Beat Analysis App. Last Accessed 26 Feb 2024. [Online]. Available: https://apps.garmin.com/apps/c82492f9-153d-4bc6-bc0d-65ad1e418f48?tid=2
55. Elite HRV. Last Accessed 26 Feb 2024. [Online]. Available: https://elitehrv.com/
56. Weltory App. Last Accessed 26 Feb 2024. [Online]. Available: https://welltory.com/
57. VARSE: VARvi Second Edition. Last Accessed: 26 Feb 2024. [Online]. Available: https://play.google.com/store/apps/details?id=com.devbaltasarq.varse
58. K. Georgiou, A.V. Larentzakis, N.N. Khamis, G.I. Alsuhaibani, Y.A. Alaska, E.J. Giallafos, Can wearable devices accurately measure heart rate variability? A systematic review. Folia Medica **60**(1), 7–20 (2018). [Online]. Available: https://www.doi.org/10.2478/folmed-2018-0012
59. T. Coppetti, A. Brauchlin, S. Müggler, A. Attinger-Toller, C. Templin, F. Schönrath, J. Hellermann, T.F. Lüscher, P. Biaggi, C.A. Wyss, Accuracy of smartphone apps for heart rate measurement. Eur. J. Prevent. Cardiol. **24**(12), 1287–1293 (2017). [Online]. Available: https://doi.org/10.1177/2047487317702044
60. M. Moya-Ramon, M. Mateo-March, I. Peña-González, M. Zabala, A. Javaloyes, Validity and reliability of different smartphones applications to measure HRV during short and ultra-short measurements in elite athletes. Comput. Methods Progr. Biomed. **217**, 106696 (2022). [Online]. Available: https://doi.org/10.1016/j.cmpb.2022.106696
61. A.T. Himariotis, K.F. Coffey, S.E. Noel, D.J. Cornell, Validity of a smartphone application in calculating measures of heart rate variability. Sensors **22**(24), 9883 (2022). [Online]. Available: https://doi.org/10.3390/s22249883
62. P. Cuesta-Morales, B.G. Perez-Schofield, L. Rodríguez-Linares, M.J. Lado, A.J. Méndez, X.A. Vila, VARSE: android app for real-time acquisition and analysis of heart rate signals. Int. J. Med. Inf. **160**, 104692 (2022). Available: https://doi.org/10.1016/j.ijmedinf.2022.104692
63. A.N. Vest, G. Da Poian, Q. Li, C. Liu, S. Nemati, A.J. Shah, G.D. Clifford, An open source benchmarked toolbox for cardiovascular waveform and interval analysis. Physiol. Measur. **39**(10), 105004 (2018). [Online]. Available: https://doi.org/10.1088/1361-6579/aae021
64. L. Rodríguez-Liñares, A.J. Méndez, M.J. Lado, D.N. Olivieri, X. Vila, I. Gómez-Conde, An open source tool for heart rate variability spectral analysis. Comput. Methods Progr. Biomed. **103**(1), 39–50 (2011). [Online]. Available: https://doi.org/10.1016/j.cmpb.2010.05.012
65. The RHRV Repository. Last Accessed 26 Feb 2024. [Online]. Available: https://rhrv.r-forge.r-project.org/
66. HRVAS. Last Accessed: 26 Feb 2024. [Online]. Available: https://github.com/jramshur/HRVAS
67. T. Kaufmann, S. Sütterlin, S.M. Schulz, C. Vögele, ARTiiFACT: a tool for heart rate artifact processing and heart rate variability analysis. Behavior Res. Methods **43**, 1161–1170 (2011). [Online]. Available: https://doi.org/10.3758/s13428-011-0107-7
68. ARTiiFACT. Last Accessed 26 Feb 2024. [Online]. Available: https://github.com/tobias-kaufmann/ARTiiFACT
69. P. Perakakis, M. Joffily, M. Taylor, P. Guerra, J. Vila, KARDIA: a Matlab software for the analysis of cardiac interbeat intervals. Comput. Methods Progr. Biomed. **98**(1), 83–89 (2010). [Online]. Available: https://doi.org/10.1016/j.cmpb.2009.10.002
70. 1 KARDIA. Last Accessed 26 Feb 2024. [Online]. Available: https://github.com/perakakis/KARDIA
71. R. Bartels, L. Neumamm, T. Peçanha, A.R.S. Carvalho, SinusCor: an advanced tool for heart rate variability analysis. Biomed. Eng. Online **16**, 1–15 (2017). [Online]. Available: https://doi.org/10.1186/s12938-017-0401-4
72. SinusCor. Last Accessed 26 Feb 2024. [Online]. Available: http://rhenanbartels.github.io/SinusCor/

References

73. L. Rodríguez-Liñares, M.J. Lado, X. Vila, A.J. Méndez, P. Cuesta, gHRV: Heart rate variability analysis made easy. Comput. Methods Progr. Biomed. **116**(1), 26–38 (2014). [Online]. Available: https://doi.org/10.1016/j.cmpb.2014.04.007
74. gHRV. Last Accessed 26 Feb 2024. [Online]. Available: https://milegroup.github.io/ghrv/
75. M. McConnell, B. Schwerin, S. So, B. Richards, RR-APET-Heart rate variability analysis software. Comput. Methods Progr. Biomed. **185**, 105127 (2020). [Online]. Available: https://doi.org/10.1016/j.cmpb.2019.105127
76. RR-APET. Last Accessed 26 Feb 2024. [Online]. Available: https://gitlab.com/MegMcC/rr-apet-hrv-analysis-software
77. M.P. Tarvainen, J.-P. Niskanen, J.A. Lipponen, P.O. Ranta-Aho, P.A. Karjalainen, Kubios HRV–heart rate variability analysis software. Comput. Methods Progr. Biomed. **113**(1), 210–220 (2014). [Online]. Available: https://doi.org/10.1016/j.cmpb.2013.07.024
78. Kubios Lite. Last Accessed 26 Feb 2024. [Online]. Available: https://www.kubios.com/hrv-scientific-lite/
79. Nevorkard. Last Accessed 26 Feb 2024. [Online]. Available: http://www.nevrokard.eu/maini/hrv.html
80. LabVIEW. Last Accessed 26 Feb 2024. [Online]. Available: https://www.ni.com/docs/en-US/bundle/labview-biomedical-toolkit-api-ref/page/lvbiomed/bio_hrv_vis.html
81. BIOPAC Systems. Last Accessed 26 Feb 2024. [Online]. Available: https://www.biopac.com/product/heart-rate-variability-analysis-software/

Chapter 2
Loading, Plotting, and Filtering RR Intervals

Abstract The initial steps to work with RHRV functions are presented in this chapter. The process starts with the loading of records containing beat positions, which should be preprocessed prior to frequency, time, or nonlinear analysis. Data can be stored in various types of files, and RHRV routines can deal with different data formats. Next, heart rate must be obtained from beat positions. It may occur that spurious points appear in the heart rate signal. RHRV allows users to delete these outliers, when necessary. Besides, the signal can be filtered to reject automatically points that do not correspond to acceptable physiological values.

2.1 Getting Started

RHRV is a free software package to analyze HRV in frequency, time, and nonlinear domains. The starting point is a beat position series that must be loaded into a specific structure, described later in the text. RHRV is not able to process ECG directly. In this case, before starting working with RHRV, it is necessary to use a tool capable of identifying the beats and exporting them to any of the multiple formats supported by RHRV (interested readers can check Appendix B).

Once loaded into RHRV, data containing beat positions must be processed to obtain a heart rate signal that can be used to perform different analyses. In order to obtain satisfactory results, outliers should be first removed (manually or automatically) and data must be interpolated.

Figure 2.1 summarizes this process, including the RHRV functions that should be used. The rest of this chapter explains all these functionalities in depth.

2.2 Data File Format

To perform HRV analysis from the sequence of beat positions, it will be necessary to generate the corresponding heart rate signal. This signal can then be analyzed in terms of frequency, time, or nonlinear parameters.

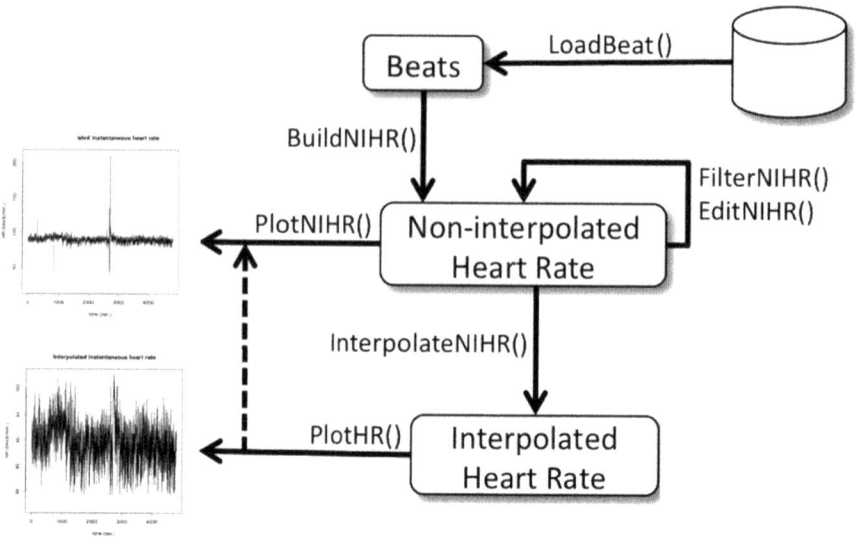

Fig. 2.1 Steps of the preprocessing procedure to obtain the heart rate signal

In RHRV, data acquisition can be performed from a simple American Standard Code for Information Interchange (ASCII) file, that is, a single-column file containing beat positions, which may appear in different units (seconds, milliseconds, etc.). However, data are not always available in this format, thus implying a previous conversion, or using alternative functions to load it into RHRV. In the following paragraphs, we will describe the most commonly used formats to store beat position information.

The WFDB (WaveForm DataBase) software is a collection of tools for viewing, analyzing, and creating recordings of physiological signals. These tools are freely distributed from the PhysioNet Website [1]. We will use the term "WFDB format" to refer to those records that follow the standard accepted by the WFDB software tools, which is one of the most used formats by researchers in ECG signal processing.

The WFDB format uses header files to specify the format and attributes of signal and annotation files. Although the WFDB format can be used to store records of different physiological signals, it is mainly used to store ECG records. In this case, the header contains information about the number of leads, calibration format, etc. ECG samples are stored as binary files, and, if there is information about beat positions or other annotations, it will be stored in one or more annotation files.

Nowadays, WFDB is being widely used by a considerable part of the research community, as well as by different organizations. However, there are many other format files worth presenting here. EDF+ (European Data Format) is another extended format, characterized by its flexibility and simplicity [2, 3]. It is commonly used for exchange and storage of multichannel biological and physical signals, which are recorded in .edf files.

EDF+ is based on EDF [4], but it can store many more types of data. It is a binary file with two main parts: header, which identifies the record and specifies the technical characteristics of the stored signals, and data, corresponding to the samples of the signals.

Improvements in technology have enabled the development of an increasing number of mobile devices (smartphones, tablets, watches, etc.), which can be used for obtaining real-time heart rate data by means of a chest strap. Products can be found for sport applications, such as Sports Tracker [5], Runtastic Heart Rate Monitor [6], Cardiio [7], or Endomondo [8]. Most of these applications obtain and show only the mean heart rate, but some recent chest straps are also able to detect instantaneous beat positions and to store and export these values. One of these is the Polar WearLink band, which can record the heart rate in a very simple and easy way using a specific data format [9], or the Apple Watch, which allows not only measuring the heart rate but also taking an electrocardiogram with the corresponding app [10].

Some Polar devices generate a .hrm file, storing data in ASCII format. Each file is composed of several data sections that are separated by empty lines. In each line, the data section name is separated from the data by using brackets. Multiple data in each row are separated by tab characters, including optional information about the exercise, such as speed, cadence, and altitude.

Another widely used format is the one used by Suunto (.sdf files), one of the most important watch manufacturers in the field of trekking, scuba diving, or climbing [11]. These .sdf files also store information in ASCII format and contain several sections delimited by their names between brackets. A header section specifies all the parameters that must be considered, each of them in one line, if they are present in the file (not all of them are mandatory). Data is stored in other sections, separating the different attributes by colons.

Apart from these most commonly used formats, data can be also stored in RR files, which are usually text files that only include the time differences between each consecutive beats.

In the next section, we will explain how to load data into RHRV from different file formats, namely, ASCII, WFDB, EDF+, Polar, Suunto, and RR formats.

2.3 Loading Beat Series into RHRV

In order to load the heartbeat positions from the data files, a custom data structure, called *HRVData*, must be initially created by using the RHRV software package. Figure 2.2 shows the fields contained in *HRVData*. This structure is implemented in R language as a list data object, and it will store in different fields all the information related to the digital records, including date and time (*datetime* field), verbose (returning of information to the user, *Verbose* field), beat positions (*Beat* field), interpolated heart rate series (*HR* field), frequency of interpolation (*Freq_HR* field), episodes (specific intervals within the heartbeat time series that can exist or

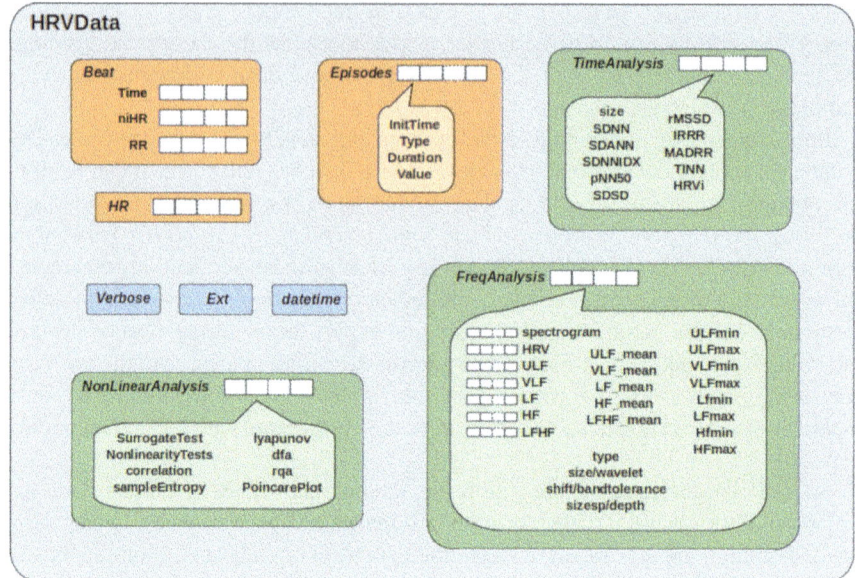

Fig. 2.2 *HRVData* structure

not, *Episodes* field), and parameters resulting from the different time (*TimeAnalysis* field), frequency (*FreqAnalysis* field), and nonlinear analyses (*NonLinearAnalysis* field).

All the fields in this structure will be carefully described in this book. Meanwhile, in this paragraph, we will explain the fields of the data structure needed to load and preprocess the original data:

- *datetime*: information related to the date and time associated with the original record, provided in the original data file or specified by the user.
- *Verbose*: Boolean argument to enable or disable verbose mode.
- *Beat*: dataframe to store information on the instantaneous heart rate data. It contains the original beat positions (*Time*), the instantaneous, non-interpolated frequency (*niHR*), and the RR intervals (distances between two consecutive beats, *RR*).
- *HR*: vector containing the interpolated heart rate data.
- *Freq_HR*: the heart rate interpolation frequency.
- *Ext*: string used as file extension by loading/writing functions.

It is important to notice that some fields of the *HRVData* structure may be empty. These fields will be filled by processing routines when applied over the original heartbeat position signal.

Listing 2.3.1 shows the code to create the *HRVData* structure.

2.3 Loading Beat Series into RHRV

R listing 2.3.1

```
# HRVData structure containing the heart beats
library(RHRV)
hrv.data <- CreateHRVData()
hrv.data <- SetVerbose(hrv.data, TRUE)
```

Once the data structure has been created, the beats should be loaded. Depending on the original file format, the user must employ a different R routine. We will show how to use various loading routines in the following paragraphs.

Different functions can be used to load beat data into the *HRVData* structure:

- *LoadBeatAscii*: for data stored in an ASCII file
- *LoadBeatWFDB*: for WFDB format records
- *LoadBeatEDFPlus*: for EDF+ data files

However, a more general function (*LoadBeat*) can be used to load records in the RHRV software package. In this case, the format should be specified as an argument of this function.

As a practical example, we will consider now a record stored in the WFDB format, that can be freely downloaded from PhysioNet [12]. It belongs to the European ST-T Database, intended to be used in the evaluation of algorithms for the analysis of ST and T-wave changes [13, 14]. In the example, we are going to use the "e0115" file (header, data, and annotation files).

As previously said, to load the beat position information, the user can employ either the generic *LoadBeat* or the *LoadWFDB* functions. In both cases, the *LoadHeaderWFDB* function, which reads the header file in order to obtain information, is implicitly executed. In Listing 2.3.2, the code to load the original data in two different *HRVData* structures, employing both functions, is presented. Note that an argument giving the extension of the annotation file must also be provided (*annotator*, by default "qrs").

R listing 2.3.2

```
# Loading the beats in WFDB format
hrv.data1 <- CreateHRVData()
hrv.data <- SetVerbose(hrv.data1, TRUE)
hrv.data2 <- CreateHRVData()
hrv.data <- SetVerbose(hrv.data2, TRUE)
hrv.data1 <- LoadBeat("WFDB", hrv.data1,"sampleData/e0115", annotator = "atr")
hrv.data2 <- LoadBeatWFDB(hrv.data2, "sampleData/e0115", annotator = "atr")
```

As it can be observed, both routines yield the same original data, stored in *hrv.data1* and *hrv.data2*.

We will explain now how to load data in ASCII format. An example of an ASCII file that can be freely downloaded is *beat_ascii.txt*, available from the MILE Group webpage [15] and the GitHub repository that contains the code used in this book [16]. This file contains heartbeat positions, and it was generated employing the VARVI (Variability of the heArt Rate in response to Visual stImuli) software package, a free software tool developed to perform heart rate variability analysis in response to different visual stimuli [17]. In this case, the code to load the original data *beat_ascii.txt* and its result are given in Listing 2.3.3.

R listing 2.3.3

```
# Loading the beats in ASCII format
hrv.data3 <- CreateHRVData()
hrv.data <- SetVerbose(hrv.data3, TRUE)
hrv.data3 <- LoadBeatAscii(hrv.data3, "sampleData/beat_ascii.txt")
```

The same result could be obtained using *LoadBeat("Ascii", hrv.data3, "beat_ascii.txt")*.

The rest of the file formats supported by RHRV can be loaded in a similar way by using the functions *LoadBeatEDFPlus*, *LoadBeatSuunto*, *LoadBeatPolar*, and *LoadBeatRR*.

Both the generic *LoadBeat* function and the rest of the functions for loading the various data file formats can include several input parameters, such as the record path (*RecordPath*, by default ".") if the file to load is not in the working directory; the scale, factor applied in RR or ASCII format if data are not in seconds (*scale*, by default 1); the datetime, also for ASCII and RR files (*datetime*, by default "1/1/1900 0:0:0"); and the annotation type for EDF+ format (*annotationType*, by default "QRS"). For ASCII files, the possibility of loading only a specific portion of the record is also controlled by two other parameters: *starttime* and *endtime*. If they are not specified, the full record is loaded. When specific values (in seconds) are provided, only the corresponding portion of the record is loaded into the *HRVData* structure.

2.4 Preprocessing

At this point, a heartbeat sequence has been loaded into the *HRVData* structure, and it is time to preprocess the signal in order to obtain the heart rate signal. This is the starting point to perform time, frequency, and nonlinear analysis. A detailed explanation is given in the following sections.

2.4.1 Instantaneous Heart Rate Signal Extraction

The distance between two consecutive beats is the well-known RR interval [18], which corresponds to the distance of the R waves associated with each beat. The instantaneous heart rate can be defined as the inverse of the time separation between

2.4 Preprocessing

two consecutive heartbeats. In RHRV, this corresponds to the calculation of the series:

$$RR[i] = (Time[i] - Time[i-1]) * 1000$$

$$HR(i) = \frac{1000}{RR[i] * 60}$$

where *Time[i]* is the time when the beat i occurs, measured in seconds; *RR[i]* is the beat-to-beat distance, measured in milliseconds; and *HR(i)* is the instantaneous heart rate, measured in beats per minute.

2.4.2 Removing Artifacts

Algorithms to detect and classify heartbeats from ECG signals often fail or may yield incorrect outputs. This means that the instantaneous heart rate signal that is directly obtained from the RR intervals can be corrupted with artifacts, that is, undesired information originated by either extern sources or physiological events. Some QRS complexes can be missed; some anomalous, ectopic beats can be incorrectly classified as normal; or the original signal can be affected by noise. Ectopic beats are not triggered as the result of the heart rate control mechanisms, but they arise from the automatism that heart muscle fibers present. Given that the goal of HRV analysis is the study of the heart rate control mechanisms, these beats must be removed before performing a HRV analysis. Besides, as the beats are not regularly originated, the heart rate is not constant, which means that the signal will be not sampled regularly in time.

Several scientific works have tried to establish if the existing detection algorithms are adequate to perform heart rate variability analysis, or if a manual revision of detection is required [19, 20]. Conclusions of those works indicate that errors may significantly affect some variability indices, specifically those corresponding to spectral analysis [21–24].

Researchers have proposed removing artifacts employing a broad variety of algorithms. Some of these works include wavelet-based methods [25, 26], different filters and thresholding values [27, 28], or independent component analysis techniques [29, 30].

The RHRV software package provides an algorithm to automatically remove artifacts. This algorithm uses adaptive thresholding for rejecting beats whose RR value differs from previous and following beats and from a mobile mean more than a threshold value. The filter also removes points that are not within acceptable physiological values [31]. This algorithm is presented as pseudo-code in Listing 2.4.1.

R listing 2.4.1

```
ULAST = 13
LONG = 50
UMEAN = 1.5 * ULAST
MINIMUM = 12
MAXIMUM = 20
MINBMP = 25
MAXBPM = 200
from (i = 2 to NBEATS) {
  if (i < LONG)
    MEAN = mean of previous beats
  else
    MEAN = mean of the last long beats
  if ((100 * abs(hr(i) { hr(i - 1)) / hr(i - 1)) < ULAST) ||
      (100 * abs(hr(i) { hr(i + 1)) / hr(i - 1)) < ULAST) ||
      (100 * abs(hr(i) { MEAN) / MEAN) < UMEAN) &&
      (MINBPM <= hr(i) <= MAXBPM)) {
    valid beat
    SIGNALDEV = 10 + SIGNALDEV(last LONG beats)
    if (SIGNALDEV < MINIMUM)
      SIGNALDEV = MINIMUM
    if (SIGNALDEV > MAXIMUM)
      SIGNALDEV = MAXIMUM
    update ULAST = SIGNALDEV
    update UMEAN = 1.5 * SIGNALDEV
  }
  else
    i = i + 1
}
```

This algorithm uses as a starting point some prefixed values (experimentally calculated) that include the initial threshold values for the filter (*ULAST = 13* and *UMEAN = 1.5 * ulast*), the number of beats (*LONG*) employed to calculate the mean value and corresponding standard deviation, and minimum (*MINIMUM = 12*) and maximum (*MAXIMUM = 20*) values allowed for the *ULAST* threshold. A fixed quantity is also added to the signal deviation (*SIGNALDEV*) to obtain the adaptive threshold employed by the filter. Accepted physiological minimum (*MINBPM*) and maximum (*MAXBPM*) values are also fixed.

Using a loop, all the consecutive beats are analyzed, excluding the first and last beat, since it is not possible to compare them with the previous or next beats, respectively. The mean value of the beats is calculated considering the *LONG* previous beats, if possible. A beat will be considered as a valid one if the corresponding heart rate differs from the previous or from the next beat less than *ULAST* (in %), or from the mean less than *UMEAN* (in %).

Threshold values are updated after a number of beats equal to *LONG*. If an artifact is detected, the next beat will be discarded, since its associated RR distance is not valid.

Most artifacts are removed by this algorithm. However, RHRV also supplies a function that provides an intuitive method to manually remove the remaining artifacts, by selecting them from a graphical window.

After the automatic and/or manual edition of the artifacts, the non-interpolated, non-equally spaced heart rate signal is obtained as a result.

2.4.3 Interpolation of the Heart Rate Signal

The non-interpolated heart rate signal obtained in the previous step can be used to obtain time and nonlinear measurements. However, in the frequency analysis, several problems may arise, since one of the main drawbacks is that sometimes the continuity of the signal is broken, and in the frequency domain, this can lead to erroneous results. Because of this, an important aspect that should be carefully studied is handling artifacts and non-interpolated signals. Furthermore, the RR signal does not have equidistant sampling, and most of the frequency algorithms require an evenly interpolated signal.

Many different interpolation algorithms can be found in the literature. The most usual are linear or cubic splines interpolation, but many others can be used. Some researchers performed interpolation using the K-nearest neighbor (k-NN) algorithm [32] or the Fourier-based methods to interpolate data [33].

When using RHRV, users can select either linear or cubic spline interpolation [34]. The sampling frequency to obtain the equally spaced heart rate signal has to be set in both cases (default is 4 Hz). After this step, a new heart rate signal that is adequate for performing spectral analysis, as well as time and nonlinear studies, is obtained.

2.5 Preprocessing Beat Data with RHRV

In this section, a detailed explanation about the calculation of RR intervals with RHRV will be provided. Filtering and edition of the heart rate signal will also be explained in a practical way.

As it was explained in former paragraphs, the heart rate signal needs to be obtained from the heartbeat data before performing different types of analysis. First, the non-interpolated heart rate signal is obtained and must be filtered to remove outliers and detection errors.

Let us consider the *hrv.data3* structure, containing the heartbeat data, created in Listing 2.3.3. From this signal, the instantaneous heart rate will be now obtained with the *BuildNIHR* function. The corresponding R code developed in the RHRV software package and results are given in Listing 2.5.1.

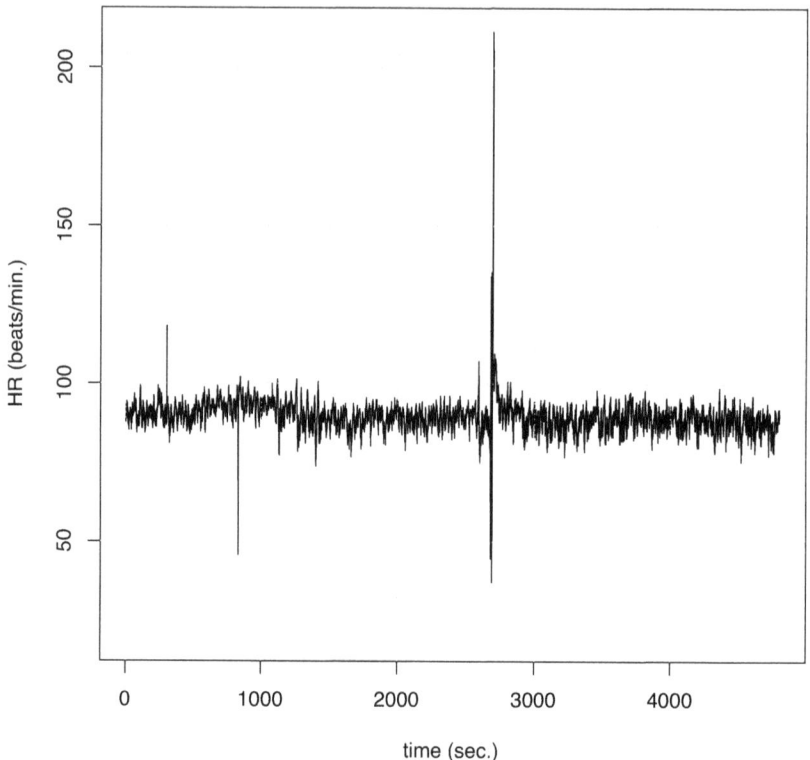

Fig. 2.3 Instantaneous heart rate signal for the "beat_ascii.txt" data

R listing 2.5.1

```
# Building the non-interpolated heart rate signal
hrv.data3 <- BuildNIHR(hrv.data3)
```

Now, the instantaneous, non-interpolated heart rate signal has been calculated. The previously empty fields in *hrv.data3* corresponding to *Beat*, that is, *niHR* and *RR*, are now filled with the appropriate values. A graphical representation can also be obtained using the function *PlotNIHR(hrv.data3)*, a plot function that allows the representation of data values for the instantaneous heart rate (Fig. 2.3).

The automatic filtering algorithm can now be applied over the heart rate signal, employing the *FilterNIHR* function. Listing 2.5.2 provides the R code and results of this function.

2.5 Preprocessing Beat Data with RHRV

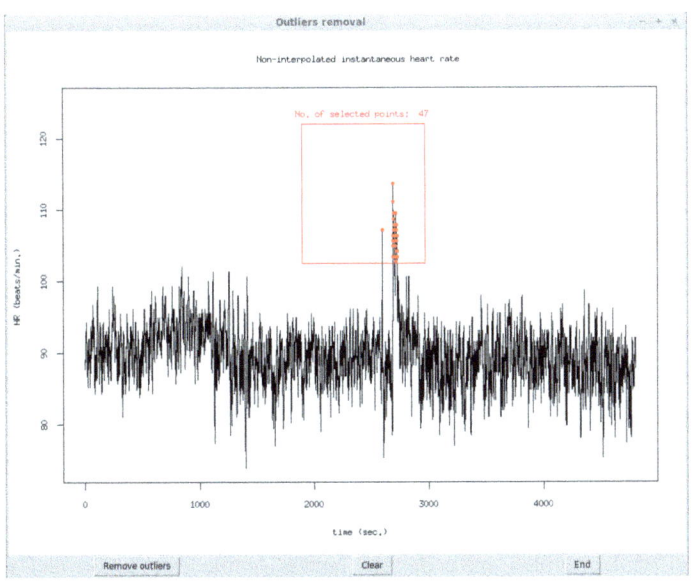

Fig. 2.4 Example of use of the *EditNIHR* function. Outliers can be manually removed

R listing 2.5.2

```
# Filtering the non-interpolated heart rate signal
hrv.data3 <- FilterNIHR(hrv.data3)
```

When the function *FilterNIHR* is used, the algorithm previously described will reject those beats that differ a certain value from the established threshold. If not specified, default parameters are used (*long* = 50 and *last* = 13).

This filtering also eliminates points that are not within acceptable physiological values. Users can modify these physiological values as well, by varying the values of the arguments *minbpm* (minimum value) and *maxbpm* (maximum value), set by default to 25 and 200, respectively. The rest of the arguments (*mini, maxi, fixed*, and *verbose*) are deprecated ones and fixed to NULL by default.

For manually removing outliers, the *EditNIHR* function can be used (Listing 2.5.3). Figure 2.4 shows an example of the use of the *EditNIHR* function. The user can interact with the graphical window and remove outliers. The user has to select the points to be removed by clicking on the upper left corner and on the bottom right corner of the area enclosing the points. This way, a red rectangular area will be marked, and the user can decide to remove the points within (Remove outliers button), to clear the selected area (Clear button), or to finish the manual edition. After removal, RHRV gives information about the removed points.

R listing 2.5.3

```
# Manual filtering the non-interpolated heart rate signal
hrv.data3 <- EditNIHR(hrv.data3)
```

In case the user does not want to use the *EditNIHR* function, but wants to remove more spurious points, the function *FilterNIHR* can be applied more than once. This can result in the elimination of more beats because, after running once the function and eliminating some of the beats, the dynamic thresholds of the filter will change. Results can be similar to the application of the *EditNIHR* function.

It is time now to estimate the interpolated heart rate signal. The function that performs this task is *InterpolateNIHR*, with the following arguments:

- Sampling frequency (*freqhr*): in Hz (default value is 4).
- Interpolation method (*method*): default value is linear, although spline can be selected.
- *Verbose*: deprecated.

Listing 2.5.4 provides the code and results of the calculation and graphical representation of the results.

R listing 2.5.4

```
#Interpolation of the instantaneous heart rate
hrv.data3 <- InterpolateNIHR(hrv.data3)
```

After applying these functions, the *HRVData* structure will contain in its fields the following: the original heartbeat positions, the RR intervals, the non-interpolated heart rate signal, and the equally spaced heart rate signal. This last field can be graphically represented employing the *PlotHR* function, shown in Listing 2.5.5 (Fig. 2.5).

R listing 2.5.5

```
# Plotting the interpolated heart rate signal
PlotHR(hrv.data3, Tags = NULL, Indexes = NULL,
       main = "Interpolated instantaneous heart rate", xlab = "time (sec.)",
       ylab = "HR (beats/min.)", type = "l", ylim = NULL, Tag = NULL,
       verbose = NULL)
```

The interpolated heart rate signal will be the starting point for the time, frequency, and nonlinear analysis, which will be explained in the next chapters of this book.

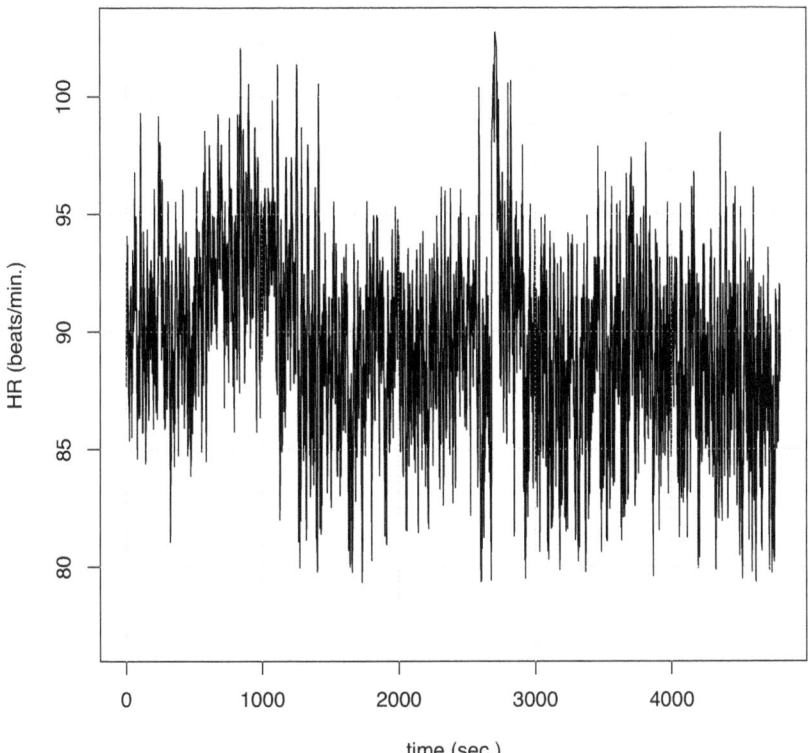

Fig. 2.5 Interpolated heart rate signal

References

1. PhysioNet. Last Accessed 26 Feb 2024. [Online]. Available: http://www.physionet.org
2. EDF+. Last Accessed 26 Feb 2024. [Online]. Available: http://www.edfplus.info
3. B. Kemp, J. Olivan, European data format 'plus'(EDF+), an EDF alike standard format for the exchange of physiological data. Clin. Neurophysiol. **114**(9), 1755–1761 (2003). [Online]. Available: https://doi.org/10.1016/S1388-2457(03)00123-8
4. B. Kemp, A. Värri, A.C. Rosa, K.D. Nielsen, J. Gade, A simple format for exchange of digitized polygraphic recordings. Electroencephalogr. Clin. Neurophysiol. **82**(5), 391–393 (1992). [Online]. Available: https://doi.org/10.1016/0013-4694(92)90009-7
5. SportsTracker. Last Accessed 26 Feb 2024. [Online]. Available: http://www.sports-tracker.com
6. Runtastic. Last Accessed 26 Feb 2024. [Online]. Available: http://www.runtastic.com
7. Cardiio. Last Accessed 26 Feb 2024. [Online]. Available: http://www.cardiio.com
8. Endomondo. Last Accessed 26 Feb 2024. [Online]. Available: http://www.endomondo.com
9. Polar. Last Accessed 26 Feb 2024. [Online]. Available: http://www.polar.com
10. Apple Watch. Last Accessed: 26 Feb 2024. [Online]. Available: https://support.apple.com/en-ca/HT204666

11. Suunto. Last Accessed 26 Feb 2024. [Online]. Available: http://www.suunto.com
12. European ST-T Database. Last Accessed 26 Feb 2024. [Online]. Available: http://www.physionet.org/physiobank/database/edb
13. A.L. Goldberger, L.A. Amaral, L. Glass, J.M. Hausdorff, P.C. Ivanov, R.G. Mark, J.E. Mietus, G.B. Moody, C.-K. Peng, H.E. Stanley, Physiobank, physiotoolkit, and physionet. Components of a new research resource for complex physiologic signals. Circulation **101**(23), e215–e220 (2000). [Online]. Available: https://doi.org/10.1161/01.CIR.101.23.e215
14. A. Taddei, G. Distante, M. Emdin, P. Pisani, G. Moody, C. Zeelenberg, C. Marchesi, The European ST-T database: standard for evaluating systems for the analysis of ST-T changes in ambulatory electrocardiography. Eur. Heart J. **13**(9), 1164–1172 (1992). [Online]. Available: https://doi.org/10.1093/oxfordjournals.eurheartj.a060332
15. gHRV. Last Accessed 26 Feb 2024. [Online]. Available: http://milegroup.github.io/ghrv/packages.html
16. C.A. García, A. Otero, X.A. Vila, M.J. Lado, L. Rodríguez-Liñares, J.M. Presedo, A.J. Méndez, Code for the book 'Heart Rate Variability Analysis with the R package RHRV' (Springer's use R!)' (2024). https://github.com/RHRV-team/RHRVBook
17. L. Rodríguez-Liñares, P. Cuesta, R. Alonso, A. Méndez, M. Lado, X. Vila, VARVI: a software tool for analyzing the variability of the heart rate in response to visual stimuli, in *Computing in Cardiology Conference (CinC), 2013* (IEEE, Piscataway, 2013), pp. 401–404
18. M.J. Goldman, *Principles of Clinical Electrocardiography* (Lange Medical Publications, New York, 1986)
19. J.T. Bigger, J.L. Fleiss, R.C. Steinman, L.M. Rolnitzky, R.E. Kleiger, J.N. Rottman, Frequency domain measures of heart period variability and mortality after myocardial infarction. Circulation **85**(1), 164–171 (1992). [Online]. Available: https://doi.org/10.1161/01.CIR.85.1.164
20. M. Malik, T. Farrell, T. Cripps, A. Camm, Heart rate variability in relation to prognosis after myocardial infarction: selection of optimal processing techniques. Eur. Heart J. **10**(12), 1060–1074 (1989). [Online]. Available: https://doi.org/10.1093/oxfordjournals.eurheartj.a059428
21. A. Molina-Picó, D. Cuesta-Frau, P. Miró-Martínez, S. Oltra-Crespo, M. Aboy, Influence of QRS complex detection errors on entropy algorithms. Application to heart rate variability discrimination. Comput. Methods Progr. Biomed. **110**(1), 2–11 (2013). [Online]. Available: https://doi.org/10.1016/j.cmpb.2012.10.014
22. R. Thuraisingham, Preprocessing RR interval time series for heart rate variability analysis and estimates of standard deviation of RR intervals. Comput. Methods Progr. Biomed. **83**(1), 78–82 (2006). [Online]. Available: https://doi.org/10.1016/j.cmpb.2006.05.002
23. L. Lu, T. Zhu, D. Morelli, A. Creagh, Z. Liu, J. Yang, F. Liu, Y.-T. Zhang, D.A. Clifton, Uncertainties in the analysis of heart rate variability: a systematic review. IEEE Rev. Biomed. Eng. **17**, 180–196 (2024). [Online]. Available: https://doi.org/10.1109/RBME.2023.3271595
24. M.D. Peláez-Coca, A. Hernando, J. Lázaro, E. Gil, Impact of the PPG sampling rate in the pulse rate variability indices evaluating several fiducial points in different pulse waveforms. IEEE J. Biomed. Health Inf. **26**(2), 539–549 (2021). [Online]. Available: https://doi.org/10.1109/JBHI.2021.3099208
25. J. Mcnames, T. Thong, M. Aboy, Impulse rejection filter for artifact removal in spectral analysis of biomedical signals, in *Engineering in Medicine and Biology Society, 2004. IEMBS'04. 26th Annual International Conference of the IEEE*, vol. 1 (IEEE, Piscataway, 2004), pp. 145–148. [Online]. Available: https://doi.org/10.1109/IEMBS.2004.1403112
26. Z. Zidelmal, A. Amirou, M. Adnane, A. Belouchrani, QRS detection based on wavelet coefficients. Comput. Methods Progr. Biomed. **107**(3), 490–496 (2012). [Online]. Available: https://doi.org/10.1016/j.cmpb.2011.12.004
27. H. Al Osman, M. Eid, A. El Saddik, A pattern-based windowed impulse rejection filter for nonpathological HRV artifacts correction. IEEE Trans. Instrum. Meas. **64**(7), 1944–1956 (2015). [Online]. Available: https://doi.org/10.1109/TIM.2014.2370496
28. J.-P. Niskanen, M.P. Tarvainen, P.O. Ranta-Aho, P.A. Karjalainen, Software for advanced HRV analysis. Comput. Methods Progr. Biomed. **76**(1), 73–81 (2004). [Online]. Available: https://doi.org/10.1016/j.cmpb.2004.03.004

References

29. F. Gimeno-Blanes, J.L. Rojo-Álvarez, J. Requena-Carrión, E. Everss, J. Hernández-Ortega, F. Alonso-Atienza, A. García-Alberola, Denoising of heart rate variability signals during tilt test using independent component analysis and multidimensional recordings, in *Computers in Cardiology, 2007*. (IEEE, Piscataway, 2007), pp. 399–402. [Online]. Available: https://doi.org/10.1109/CIC.2007.4745506
30. M. Chitkara, Design of machine learning approach based characterization of HRV dynamics during meditation using multi-domain HRV features. SN Comput. Sci. **4**(6), 769 (2023). [Online]. Available: https://doi.org/10.1007/s42979-023-02234-w
31. J. Vila, F. Palacios, J. Presedo, M. Fernández-Delgado, P. Félix, S. Barro, Time-frequency analysis of heart-rate variability. Eng. Med. Biol. Mag. IEEE **16**(5), 119–126 (1997). [Online]. Available: https://doi.org/10.1109/51.620503
32. S. Begum, M.S. Islam, M.U. Ahmed, P. Funk, K-NN based interpolation to handle artifacts for heart rate variability analysis, in *IEEE International Symposium on Signal Processing and Information Technology (ISSPIT), 2011* (IEEE, Piscataway, 2011), pp. 387–392. [Online]. Available: https://doi.org/10.1109/ISSPIT.2011.6151593
33. Z. Wei, C. Dechang, W. Xueyun, L. Hongxing, Heart rate estimation by iterative Fourier interpolation algorithm. Electr. Lett. **50**(24), 1799–1801 (2014). [Online]. Available: https://doi.org/10.1049/el.2014.3086
34. J. Friedman, T. Hastie, R. Tibshirani, *The Elements of Statistical Learning*, vol. 2, no. 1 (Springer, Berlin, 2009). [Online]. Available: https://doi.org/10.1007/978-0-387-21606-5

Chapter 3
Time-Domain Analysis

Abstract Heart rate variability can be analyzed by several methods, being time-domain methods the simplest measures that can be obtained from RR interval series. Numerical indices summarizing the variability of the series are calculated from the RR interval series. These measures are easily implemented and have low computational cost. They are usually calculated over longer segments of data than frequency-domain methods, typically 24 h. However, they can be also estimated from smaller segments (typically 5 min) in order to compare different episodes. In this chapter, a review of time-domain measures and their calculation within RHRV will be shown.

3.1 Time-Domain Measures

Researchers have proposed an extensive number of HRV indices that can be estimated in the time domain [1–3]. The most used ones are summarized in Table 3.1.

The arithmetic mean of the RR interval series, $\overline{RR}$, can be defined as follows:

$$\overline{RR} = \frac{1}{N}\sum_{i=1}^{N} RR_i$$

Being N the length of the RR series, and RR_i the RR interval between beats i and $i-1$, where each beat position corresponds to the beat detection instant. *SDNN* is the standard deviation of the RR interval series:

$$SDNN = \sqrt{\frac{1}{N-1}\sum_{i=1}^{N}(RR_i - \overline{RR})^2}$$

SDNN reflects the global variability of the series. However, *SDNN* value depends on the length of the recording, and therefore durations of recording should be standardized in order to obtain comparable results. As discussed in [3], both short-

Table 3.1 Most important time-domain measures

Index	Units	Description
SDNN	ms	Standard deviation of RR interval series
SDANN	ms	Standard deviation of the mean of RR intervals in 5 min segments of the entire series
SDNNIDX	ms	Mean of the standard deviation of RR intervals in 5 min segments of the entire series
pNN50	%	Proportion of adjacent RR intervals differing by more than 50 ms
SDSD	ms	Standard deviation of differences between adjacent RR intervals
r-MSSD	ms	The square root of the mean of the squares of differences between adjacent RR intervals
IRRR	ms	Difference between the third and first quartile of the RR interval series
MADRR	ms	Median of the absolute differences between adjacent RR intervals
TINN	ms	Baseline width of the triangular interpolation of the interval histogram
HRV index	ms	Integral of the interval histogram divided by its maximum

term (5 min) and long-term (24 h) recordings are commonly used to calculate *SDNN* as well as other HRV measures.

Other time-domain measures are estimated from segments of the total recording, usually 5 min. For instance, *SDANN* is the standard deviation of the means of these segments, while *SDNNIDX* is the mean of the standard deviations of these same segments. To compute these indices, the RR series is divided into fragments 5 min long, and the mean and standard deviation of RR distances can be calculated for each fragment:

$$\overline{RR}_s = \frac{1}{N_s}\sum_{i=1}^{N_s} RR_i$$

$$SDNN_s = \sqrt{\frac{1}{N_s - 1}\sum_{i=1}^{N_s}(RR_i - \overline{RR}_s)^2}$$

where $\overline{RR}_s$ and $SDNN_s$ are the mean and the standard deviation, respectively, of the RR distances in fragment s and N_s is the number of elements in this fragment. As said, *SDANN* is the standard deviation of the means of all the fragments, and *SDNNIDX* is the mean of the standard deviations of all fragments:

$$\overline{RR}_{alls} = \frac{1}{m}\sum_{s=1}^{m} \overline{RR}_s$$

$$SDANN = \sqrt{\frac{1}{m-1}\sum_{s=1}^{m}(\overline{RR}_s - \overline{RR}_{alls})^2}$$

3.1 Time-Domain Measures

$$SDNNIDX = \frac{1}{m}\sum_{s=1}^{m} SDNN_s$$

where $\overline{RR}_{alls}$ is the mean of all segments and m is the total number of segments. *SDANN* evaluates long-term variability, due to cycles longer than 5 min, and *SDNNIDX* quantifies short-term variability, because of cycles shorter than 5 min.

Indices based on the *difference between adjacent RR values* quantify the power due to high-frequency components. The most used are *pNN50*, *SDSD*, and *r-MSSD*. *pNN50* is the proportion of values of the RR series that differ from the previous value more than 50 ms:

$$\Delta RR_j = RR_{j+1} - RR_j, \; for \; j = 1, \ldots, N-1$$

$$NN50 = number \; of \; \Delta RR_j > 50 \, ms$$

$$pNN50 = \frac{1}{N-1} NN50 \cdot 100$$

SDSD is the standard deviation of differences between adjacent RR intervals:

$$\overline{\Delta RR} = \frac{1}{N-1}\sum_{j=1}^{N-1} \Delta RR_j$$

$$SDSD = \sqrt{\frac{1}{N-2}\sum_{j=1}^{N-1}(\Delta RR_j - \overline{\Delta RR})^2}$$

And *r-MSSD* is defined as the square root of the mean of the squares of differences between adjacent RR intervals:

$$r - MSSD = \sqrt{\frac{1}{N-1}\sum_{j=1}^{N-1} \Delta RR_j^2}$$

These three short-term measurements estimate high-frequency variations and, therefore, show high correlation with each other. Other similar indices are *IRRR* and *MADRR*, correlated with *SDNN* and *pNN50*, respectively, and that solve some of their problems. *IRRR* is the interquartile range, that is, the difference between the third and the first quartile of the RR series, and it is a very robust measure against artifacts:

$$IRRR = Q_3(\Delta RR) - Q_1(\Delta RR)$$

Quartiles are defined as the three values of the points that divide the ordered dataset into four equally sized groups. The first quartile (Q_1) splits off the lowest 25% of the RR values, and the third quartile (Q_3) splits off the highest 25% of the RR values. If the RR series has a Gaussian distribution, *IRRR* is an estimation of standard deviation (overall HRV).

MADRR is the median of the absolute differences between adjacent RR values and does not suffer from saturation problems as *pNN50*. Similar to *pNN50*, *MADRR* measures the power due to high-frequency components. This median is calculated as the second quartile (Q_2) of $|\Delta RR|$:

$$MADRR = Q_2(|\Delta RR|)$$

In addition to these statistical parameters, there are some geometric measures that can be estimated from the RR interval histogram, such as *TINN* and the *HRV index*, in order to avoid the negative influence of artifacts in the determination of variability indices. A bin of approximately 8 ms (exactly 7.8125 ms) can be used to construct smoothed histograms [3].

TINN is the baseline width of the triangular interpolation of the RR interval histogram. *TINN* may be calculated as the total area of the histogram of the RR series, divided by the maximum of the histogram multiplied by 2:

$$TINN = \frac{N \cdot bin\ size}{max(histogram(RR))} \cdot 2$$

Figure 3.1 shows an example of the triangular interpolation of RR histogram using this method.

An approximation to this measure is the *HRV index*. The *HRV index* is the integral of the interval histogram (the number of all RR intervals, that is, the length of the RR series) divided by the maximum of the interval histogram. The *HRV index* does not need triangular interpolation and it is dimensionless:

$$HRV\ index = \frac{N}{max(histogram(RR))}$$

Both measures, *TINN* and *HRV index*, reflect overall HRV and are more influenced by the lower than by the higher frequencies.

3.2 Time-Domain Analysis with RHRV

Time-domain analysis with RHRV [4] is very simple, given that only the execution of one function is needed: *CreateTimeAnalysis*. Let us consider the *hrv.data3* structure previously created in Chap. 2. Once the signal *beat_ascii.txt* (this file can be found in the GitHub repository that contains the code used in this book [5])

3.2 Time-Domain Analysis with RHRV

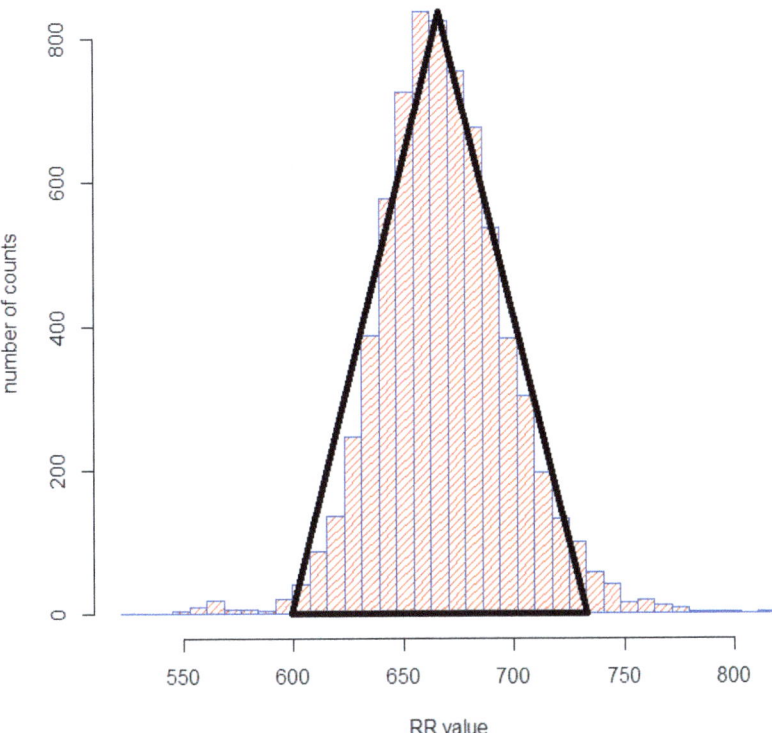

Fig. 3.1 Triangular interpolation of the RR histogram

is loaded, and the instantaneous heart rate (RR series) is calculated and filtered to eliminate artifacts, the *CreateTimeAnalysis* function modifies the *HRVData* structure to include all the time-domain measures mentioned in the previous section (Listing 3.2.1).

R listing 3.2.1

```
# Creating time analysis for the hrv.data3 structure created in chapter 2.

hrv.data3 <- CreateHRVData(Verbose = FALSE)
hrv.data3 <- LoadBeatAscii(hrv.data3, "sampleData/beat_ascii.txt")

# Building the non-interpolated heart rate signal
hrv.data3 <- BuildNIHR(hrv.data3)

# Filtering the non-interpolated heart rate signal
#(twice to eliminate all artifacts)
hrv.data3 <- FilterNIHR(hrv.data3)
hrv.data3 <- FilterNIHR(hrv.data3)
```

```
hrv.data3 <- SetVerbose(hrv.data3, TRUE)
hrv.data3 <- CreateTimeAnalysis(hrv.data3)

##   Creating time analysis
##   Size of window: 300 seconds
##   Width of bins in histogram: 7.8125 milliseconds
##   Number of windows: 15
##   Data has now 1 time analyses
##     SDNN: 30.2246 msec.
##     SDANN: 12.3131 msec.
##     SDNNIDX: 26.9153 msec.
##     pNN50: 0.4743 %
##     SDSD: 15.0925 msec.
##     r-MSSD: 15.0915 msec.
##     IRRR: 40 msec.
##     MADRR: 8 msec.
##     TINN: 133.8299 msec.
##     HRV index: 8.5651
```

It should be noted that it is not necessary to perform the interpolation process before applying the time-domain techniques, since these parameters are calculated directly from the non-interpolated RR time series. In this example, default parameters are used, although the following arguments can be used:

- *size*: length of segments in which the RR series is divided (default: 300 seconds).
- *numofbins*: number of bins in RR histogram; when not specified, the interval parameter is used (default).
- *interval*: width of bins in RR histogram (default: 7.8125 ms).

Each time-domain measure can be accessed directly in the *HRVData* structure. For instance, to access the *IRRR* value, *hrv.data3$TimeAnalysis[[1]]$IRRR* can be used. RHRV allows users to have more than one time analysis by using the *CreateTimeAnalysis* function more than once. For example, if this function were used twice, the *HRVData* structure would have another element in the *TimeAnalysis* field (*hrv.data3$TimeAnalysis[[2]]*). The *TimeAnalysis* field index is increased automatically when a new time analysis is created.

3.3 Changes in HRV Time-Based Statistics Under Pathological Conditions

Many HRV measures show correlation among them [1]. *SDNN* is an estimate of the overall HRV, such as *IRRR*, *TINN*, and *HRV index*; *SDANN* is an estimate of long-term components of HRV; and *SDNNIDX* is an estimate of short-term components of HRV, such as *pNN50*, *SDSD*, *r-MSSD*, and *MADDR*. Therefore, the following four measures are recommended for time-domain HRV assessment: *SDNN* and *HRV index* (estimate of overall HRV), *SDANN* (estimate of long-term components of HRV), and *r-MSSD* (estimate of short-term components of HRV).

Measures expressing the overall HRV and its long- and short-term components cannot replace each other, and the selection of which measures to use should correspond to the aim of each particular study. Furthermore, it is not correct to compare time-domain measures, especially those referring to the overall HRV, obtained from recordings of different durations. Although it must be taken into account that depressed HRV is generally a predictor of a nonhealthy state.

A study using RHRV was carried out with the purpose of assessing possible correlations between HRV parameters and personality impulsive traits, in order to establish if subjects are at risk of suffering from suicidal ideation [6]. The results showed a significant drop on some time-domain parameters (*MADRR*, *IRRR*, *pNN50*) from the healthy group to the patients showing more impulsivity. Despite the small number of cases, the smaller value of these parameters indicates that HRV could be used as a tool for the diagnosis of suicidal ideation.

Depressed HRV in the acute phase of myocardial infarction has also been reported, particularly the reduction in 24-h *SDNN* [7]. Furthermore, in neuropathy associated with diabetes mellitus, a reduction in time-domain parameters of HRV was also observed [3]. In chronic mitral regurgitation, *SDANN* correlated well with ventricular performance and predicted clinical events [8]. *SDNN* was significantly related to 1 year mortality in sudden cardiac arrest [9]. Moreover, using simple time-domain methods, about one-fourth of 107 alcoholic patients were identified as having autonomic neuropathy [10].

Additionally, heart rate variability appears to be a potential good indicator of cumulated training load in middle-distance runners, which could help to plan training programs [11].

References

1. U.R. Acharya, K.P. Joseph, N. Kannathal, C.M. Lim, J.S. Suri, Heart rate variability: a review. Med. Biol. Eng. Comput. **44**(12), 1031–1051 (2006). [Online]. Available: https://doi.org/10.1007/s11517-006-0119-0
2. M. García-González, R. Pallás-Areny, Novel, robust indexes for time-domain analysis of heart rate variability, in *Proceedings of the 18th Annual International Conference of the IEEE*, vol. 4 (1996), pp. 1660–1661. [Online]. Available: https://doi.org/10.1109/IEMBS.1996.647600
3. Task Force of the European Society of Cardiology and the North American Society of Pacing and Electrophysiology, Heart rate variability: Standards of measurement, physiological interpretation and clinical use. Eur Heart J **17**, 354–381 (1996). [Online]. Available: https://doi.org/10.1093/oxfordjournals.eurheartj.a014868
4. L. Rodríguez-Liñares, A.J. Méndez, M.J. Lado, D.N. Olivieri, X.A. Vila, I. Gómez-Conde, An open source tool for heart rate variability spectral analysis. Comput Methods Progr. Biomed **103**(1), 39–50 (2011). [Online]. Available: https://doi.org/10.1016/j.cmpb.2010.05.012
5. C.A. García, A. Otero, X.A. Vila, M.J. Lado, L. Rodríguez-Liñares, J.M. Presedo, A.J. Méndez, Code for the book 'Heart Rate Variability Analysis with the R package RHRV' (Springer's use R!) (2024). https://github.com/RHRV-team/RHRVBook .

6. A.J. Méndez, M.J. Lado, X.A. Vila, L. Rodríguez-Liñares, R.A. Alonso, A. García-Caballero, Heart of darkness: Heart rate variability on patients with risk of suicide, in *8th Iberian Conference on Information Systems and Technologies (CISTI)* (2013), pp. 1–4
7. G.C. Casolo, P. Stroder, C. Signorini, F. Calzolari, M. Zucchini, E. Balli, A. Sulla, S. Lazzerini, Heart rate variability during the acute phase of myocardial infarction. Circulation **85**(6), 2073–2079 (1992). [Online]. Available: https://doi.org/10.1161/01.CIR.85.6.2073
8. K.M. Stein, J.S. Borer, C. Hochreiter, P.M. Okin, E.M. Herrold, R.B. Devereux, P. Kligfield, Prognostic value and physiological correlates of heart rate variability in chronic severe mitral regurgitation. Circulation **88**(1), 127–135 (1993). [Online]. Available: https://doi.org/10.1161/01.CIR.88.1.127
9. C.M. Dougherty, R.L. Burr, Comparison of heart rate variability in survivors and nonsurvivors of sudden cardiac arrest. Am. J. Cardiol. **70**(4), 441–448 (1992). [Online]. Available: https://doi.org/10.1016/0002-9149(92)91187-9
10. R. Monforte, R. Estruch, J. Valls-Solé, J. Nicolás, J. Villalta, Á. Urbano-Márquez, Autonomic and peripheral neuropathies in patients with chronic alcoholism: a dose-related toxic effect of alcohol. Arch. Neurol. **52**(1), 45–51 (1995). [Online]. Available: https://doi.org/10.1001/archneur.1995.00540250049012
11. V. Pichot, F. Roche, J.-M. Gaspoz, F. Enjolras, A. Antoniadis, P. Minini, F. Costes, T. Busso, J.-R. Lacour, J.C. Barthelemy, Relation between heart rate variability and training load in middle-distance runners. Med Sci Sports Exercise **32**(10), 1729–1736 (2000). [Online]. Available: https://doi.org/10.1097/00005768-200010000-00011

Chapter 4
Frequency-Domain Analysis

Abstract The sympathetic and parasympathetic branches of the ANS strongly influence heart rate. Afferent sympathetic activity increases heart rate, while afferent parasympathetic activity decreases heart rate. The speed with which changes in these systems are reflected in changes in heart rate is different. The sympathetic system is slow in its effects (a few seconds), while the parasympathetic system is faster (0.2–0.6 s). Given the different speeds of response, it is possible to use frequency analysis to study sympathetic and parasympathetic contributions to the HRV. A key fact to keep in mind in this analysis is that the RR series is not stationary. In this chapter, we will see how to perform HRV frequency analysis using RHRV.

4.1 Frequency Components of the HRV

Akselrod et al. [1] were the first to realize that power spectrum analysis of heart rate fluctuations provides quantitative information about the function of the cardiovascular control systems. Akselrod et al. described three components in the HRV power spectrum with physiological relevance: the very-low-frequency component (VLF, frequencies below 0.03 Hz), the low-frequency component (LF, 0.03–0.15 Hz), and the high-frequency component (HF, 0.15–0.4 Hz). At present, sometimes the VLF component is split into two components: the VLF band (0.003–0.03 Hz) and the ultralow-frequency band (ULF, 0–0.003 Hz).

Although the limits given here for these frequency bands are the most used in the literature when analyzing the HRV of adult humans at rest, or while performing a low level of physical activity, there is not an absolute consensus on the precise limits of these bands' boundaries [2–5]. Moreover, these bands' boundaries are not suitable for the study of HRV in children [6] or during exercise [7]. For example, the upper limit of the HF may extend up to 1 Hz for children or for adults during exercise. These boundaries are neither suitable for the study of HRV in animals, especially small mammals such as mice or rats [8, 9]. The faster dynamics of cardiovascular regulation in these animals requires higher spectral boundaries in the study of the VLF, LF, and HF components. If a study of HRV in adults under conditions of physical exercise, in children, or in some animal is going to be conducted, it is advisable to review the scientific literature on the subject to find out what frequency bands are being used.

The HF component is believed to be of parasympathetic origin, but it also contains a strong contribution of the RSA, i.e., the heartbeat synchronization with the respiratory rhythm [10]. An increase of the HF activity occurs during cold stimulation of the face, rotational stimuli, and controlled respiration [11].

The LF component is a subject of certain controversy. Some authors believe it is of both sympathetic and parasympathetic origin [1, 12], although others have suggested that the sympathetic system predominates [13, 14]. This discrepancy is mainly due to the fact that under conditions of sympathetic excitation, there is a decrease in the absolute power of the LF band. This band also includes a component arising from the Mayer waves. Mayer waves are cyclic changes in arterial blood pressure caused by oscillations in the baroreceptor and chemoreceptor reflex control systems, which have a frequency of about 0.1 Hz.

An increase in LF occurs during mental stress, standing and moderate exercise in healthy subjects, hypotension, physical activity, and occlusion of a coronary artery or common carotid arteries in conscious dogs [11]. Both LF and HF exhibit a circadian pattern and reciprocal fluctuations, with higher values of LF in the daytime and higher values of HF at night [14].

The ratio of the spectral power in the LF and HF bands, often referred to as LF/HF, is considered a quantitative measure of the sympatho/vagal balance; higher values of LF/HF indicate a predominance of the sympathetic system, while lower values indicate a predominance of the parasympathetic system. However, some researchers disagree about the usefulness of this ratio [15].

At present, there is relatively little knowledge about the origins of the VLF and ULF components. While some authors have related the VLF with the renin-angiotensin system [16], others doubt that there is a specific physiological process attributable to these components. A problem associated with the study of the VLF and ULF is that they are strongly affected by baseline removal algorithms [11], which complicates the reproduction of the results obtained by other authors.

When performing a frequency analysis, researchers should always keep in mind the length of the recordings they are working with. If the recording has a length of between 1 and 5 min, the spectral power in the LF and HF bands can be estimated with a reasonable level of accuracy. But estimating the power in the VLF band provides dubious information if the recording lasts less than 5 min [11]. To obtain reliable information about VLF, 5 to 10 min is needed. And even longer recordings are needed to obtain reliable information about the ULF band: at least 30–60 min. If our recording is not long enough, we must be careful when interpreting the results obtained for the spectral power of bands for which the optimum recording length has not been met. As always in an experimental work, the more data we have, the more information we can extract from it. But, of course, it is not always possible to obtain more data.

4.2 Frequency Analysis Techniques

4.2.1 Frequency Analysis of Stationary Signals

The most basic frequency analysis technique of HRV is the Fourier transform, which decomposes a periodic signal into a sum of a (possibly infinite) set of sines and cosines. If the signal is not periodic (as in the case of finite signals), a periodic signal is created by replicating the original one (an operation known as *periodic extension*). The discrete implementation of the Fourier transform is called discrete Fourier transform (DFT), and its efficient implementation is called fast Fourier transform (FFT). The DFT of N points of the RR series, $\{RR_0, RR_1, \ldots, RR_{L-1}\}$ ($L \leq N$), is obtained by padding with zeros the original sequence (to obtain a signal of length N) and by the periodic extension of the padded sequence. This procedure permits the transformation of a sequence $\{RR_n\}$ of $L \leq N$ points in a sequence $\{X_k\}$ of N points given by

$$X_k = \sum_{n=0}^{L-1} RR_n \cdot e^{-2\pi k i \frac{n}{N}} \quad k = 0, 1, 2, \ldots, N-1.$$

X_k contains information about the normalized angular frequency $2\pi \frac{k}{N}$, $k = 0, 1, 2, \ldots, N-1$, present in signal $\{RR_n\}$. It must be noted that zero-padding does not alter the original set of frequencies present in $\{RR_n\}$. However, zero-padding provides a more detailed representation of the set of these frequencies.

Zero-padding can also be used to increase the efficiency of the FFT. If N is a highly composite number (e.g., $N = 2^l$, being l an integer), the FFT can be computed with $O(N \log N)$ operations. Otherwise $O(N^2)$ operations are needed. When the length of the RR series is not highly composite, often zero-padding is used to extend the series to the next highly composite number in order to compute the FFT efficiently.

It must also be noted that after the periodic extension, there will usually exist a jump between the end of one replicate and the beginning of the next one. These jumps are usually avoided by reducing the amplitude of the time series at the beginning and toward its end. This operation is known as *tapering*.

The DFT is a complex transformation. However, in practical applications, the so-called periodogram $\mid X_k \mid^2$ is often used. The periodogram gives information about the relative strengths of the various frequencies present in the signal, and it is usually interpreted as a sample estimate of a population function called the power spectral density (PSD).

The DFT can also be applied to stationary random signals, although its interpretation in terms of frequencies is often difficult. A simpler interpretation of this transformation is given by the Wiener-Khinchin theorem, which states that the power spectral density of a stationary process is the DFT of its autocorrelation function [17].

The calculation of the periodogram through the DFT often leads to rough estimations of the population spectral density. Furthermore, the raw periodogram is not a consistent estimator of the spectral density. However, since adjacent values are asymptotically independent, a consistent estimator can be built by smoothing the raw periodogram. In the context of time series analysis, one of the most used smoothers is the Daniell kernel [18]. An alternative approach for obtaining smoother periodograms is based on the fact that most stationary time series can be approximated by an autoregressive AR(p) model (p is the order of the process). AR(p) models state that the output of the variable depends linearly on its own p previous values:

$$RR_t = c + \sum_{i=1}^{p} \phi_i RR_{t-i} + \epsilon_t,$$

where ϵ_t is a white noise process. To obtain an estimation of the periodogram, a suitable AR(p) model is fitted to the time series. The periodogram of the time series is then computed as the expected periodogram of the fitted AR model [18]. In the context of heart rate analysis, a fixed order of $p = 16$ has been suggested [19]. PSD estimation methods that require some sort of signal model prior to the calculation of the periodogram are referred to as parametric methods. In contrast, nonparametric methods do not make any assumption about the signal. An example of a nonparametric method would be the DFT with Daniell smoothers.

The main advantages of the nonparametric method are no assumption is made about the signal (and thus there is no need to verify the suitability of the chosen model), the simplicity of the algorithm employed (FFT), and its high processing speed. On the other hand, the main advantages of parametric methods are smoother spectral components, easy identification of the central frequency of each component, and accurate estimation of PSD even with a small number of samples (provided that the signal is stationary and the AR model is suitable).

All the methods presented so far require uniformly sampled data. That is, the discrete signal is assumed to have been created by measuring the value of a continuous physical signal at a constant rate (the sampling interval). However, the RR intervals are not equidistantly sampled. The most common solution to this problem is to use interpolation methods for converting the non-equidistantly sampled RR intervals in a uniformly sampled time series. Another possible solution is using PSD estimation methods specially prepared for unevenly sampled signals. The most widely used technique of this type is the Lomb-Scargle periodogram [20, 21]. The Lomb-Scargle periodogram estimates the frequencies present on the signal by applying a least squares fit of sinusoids to the data samples. Some authors have argued that the Lomb-Scargle periodogram should be the preferred HRV analysis method, since it avoids artifactual contributions that distort the high-/low-power ratio in the classical methods [22].

4.2.2 Frequency Analysis of Nonstationary Signals

All methods presented in Sect. 4.2.1 are powerful tools for analyzing signals that do not evolve in time (stationary), but they are not good tools for studying nonstationary signals: for example, the Fourier transform can determine which frequencies are present in a signal, but not when they are present (a similar reasoning applies when autocorrelation changes in time). To be able to study nonstationary signals, several techniques capable of representing a signal in both time and frequency domain have been developed. The key idea behind these time-frequency joint representations is to define elementary time-frequency atoms as waveforms with minimum spread in the time-frequency plane [23]. Then, these time-frequency atoms (instead of the infinite time spread sines and cosines) are used to decompose the signals. Selecting the time-frequency atoms is not a trivial problem because of the time-frequency uncertainty principle: the energy spread of a function and its Fourier transform cannot simultaneously be arbitrarily small [23].

One of the most common time-frequency decompositions for discrete signals is the short-time Fourier transform (STFT), which uses a symmetric window to select an interval of the signal that is going to be analyzed. The remaining signal is analyzed by repeatedly shifting the window in time. Typically, these shifts are arranged to ensure a certain degree of overlap between the multiple time windows employed. In order to sample the frequency interval with M equally spaced frequencies, the practical implementation of the STFT of the RR series is given by

$$X_{r,k} = \sum_{n=0}^{L-1} RR_{rR+n} \cdot w_n e^{-2\pi ki \frac{n}{M}}$$

where r and k are integers fulfilling $0 \leq r \leq \left\lfloor \frac{N-L-1}{R} \right\rfloor$, N is the length of x, and $0 \leq k \leq M-1$, and w is the L-point discrete window used in the analysis whose position moves in jumps of R samples in time. To compute the STFT efficiently, the width of window w is typically chosen as a highly composite number. Using a window with a small width provides good temporal resolution, at the cost of losing frequency resolution, especially in the lower frequencies. On the contrary, a wider window provides better spectral resolution at the expense of time resolution. Although it will not be exposed here, this idea of windowing can also be applied to parametric PSD estimation.

When HRV frequency analysis is performed using the STFT, there is an inherent tension between studying short-duration phenomena (e.g., apnea) that may be reflected in the higher-frequency spectral bands (LF and HF) and obtaining information on the lower-frequency bands (VLF and ULF). If we want information about the VLF band, we should use at least a 5–10 min window. To obtain a reliable estimation about the ULF band, we should use an even wider window. But if we use very wide windows, we will not be able to capture the effect of a short-duration phenomenon, such as an apnea, that usually lasts about 30–60 s. A common

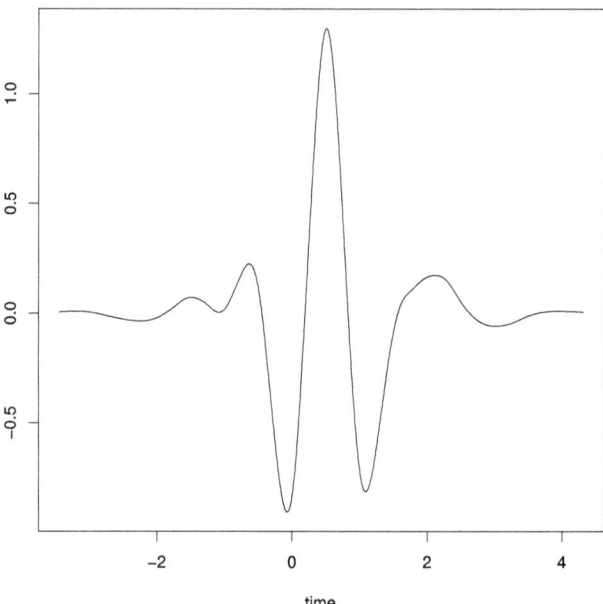

Fig. 4.1 Example of a mother wavelet

practice in these scenarios is to perform two frequency analyses. The first one uses a wide window that provides reliable information on the VLF and ULF bands, but it is not capable of extracting information about short-duration phenomenon. The second analysis uses a narrow window that can capture the effect of short-duration phenomenon on the LF and HF bands, but it does not provide a good estimation of the VLF and ULF bands.

A better solution to this problem is to use the wavelet Transform (WT). A wavelet is a "small wave" with zero mean that grows and decays in a limited time period (see Fig. 4.1). A function fulfilling these conditions is called "mother wavelet." The mother wavelet ψ can be scaled and shifted in time, yielding a set of wavelet functions with different sizes and centered in different time positions:

$$\psi_{u,s}(t) = \frac{1}{\sqrt{s}} \psi\left(\frac{t-u}{s}\right)$$

where s is positive and defines the scale and u is any real number defining the shift. This set of functions is used to extract time-frequency information by correlating them with the signal $f(t)$ being analyzed:

$$[W_\psi f](u,s) = \frac{1}{\sqrt{s}} \int_{-\infty}^{\infty} \psi\left(\frac{t-u}{s}\right) f(t) dt$$

4.3 Frequency-Domain Analysis with RHRV

where $[W_\psi f](u,s)$ is the WT of the signal $f(t)$ using the mother wavelet ψ. When working with discrete signals, and if the proper set of orthonormal wavelets is used in the decomposition, it is possible to efficiently compute the WT as a set of conjugate mirror filters and subsampling operations [24].

The WT provides a better compromise between time and frequency resolution than the STFT. The STFT uses just one window to analyze the signal. However, the ideal approximation would be to use narrow windows at high frequencies and wide windows at low frequencies. The wavelet transform uses multiple windows, obtained by applying dilatations to the mother wavelet, leading to a multiresolution analysis.

In the literature, there are more HRV studies that use the STFT than studies that use the WT, despite the theoretically superior properties of the latter. We believe that the main reason for this is the historical lack of HRV analysis tools with support for spectral analysis based on WT and the greater complexity of implementing manually these algorithms. RHRV can perform HRV frequency analysis using both techniques, and they both have similar computational efficiency. When we perform a spectral analysis with RHRV, we often use both techniques and compare the results (after all, using both techniques means adding a single line of code to the analysis script). Our experience is that, as a general rule, the WT analysis is more sensitive in detecting subtle HRV changes [24].

4.3 Frequency-Domain Analysis with RHRV

4.3.1 Frequency Analysis of Stationary Signals

To illustrate the frequency-domain analysis techniques of stationary signals, we shall briefly explore the interesting phenomena of RSA. RSA produces a shortening in RR intervals during inspiration and prolonged RR intervals during expiration. The register that will be used in this section for illustration purposes may be downloaded from [25] ("M2.hea" and "M2.qrs" files). It belongs to a volunteer recorded in supine position and breathing at a fixed rate of 0.25 Hz for 10 min (thus, ULF estimates will not be accurate). The database to which the recording belongs is described in [26].

The effect of RSA in heart rhythm is evident when plotting the RR series, as shown in Fig. 4.2. The code to obtain this figure is shown in Listing 4.3.1.

R listing 4.3.1

```
hrv.data <- CreateHRVData()
hrv.data <- LoadBeatWFDB(hrv.data,RecordName = "M2",
                        RecordPath = "sampleData",
                        annotator = "qrs")
hrv.data <- BuildNIHR(hrv.data)
PlotNIHR(hrv.data, xlim = c(200, 400))
```

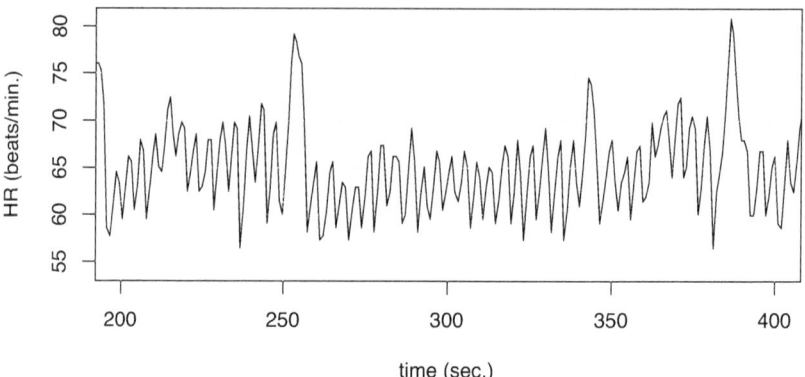

Fig. 4.2 Heart rate series from a volunteer breathing at a constant rate. Note the sinusal pattern

The sinusal pattern that can be appreciated in Fig. 4.2 should be even more evident in the frequency domain. To carry out a frequency analysis with RHRV, we must start by creating the data structure that will store the results of the analysis. This data structure stores the analysis results as a member of the *FreqAnalysis* list, under the *HRVData* structure (see Fig. 2.2). Each analysis structure created is identified by a number, starting by the number 1 for the first analysis created and increasing by 1 for every new frequency analysis created. To create the frequency analysis structure, the *CreateFreqAnalysis* function is used (see Listing 4.3.2).

R listing 4.3.2

```
hrv.data <- SetVerbose(hrv.data, TRUE)
hrv.data <- CreateFreqAnalysis(hrv.data)

## Creating frequency analysis
## Data has now 1 frequency analysis
```

If verbose mode is on, the *CreateFreqAnalysis* function prints information about the number of frequency analysis structures that have been created so far.

The main function to estimate the PSD of the RR time series is the *CalculatePSD* function. When using the *CalculatePSD* function, the user may specify the following parameters:

- *indexFreqAnalysis*: indicates in which spectral analysis structure the results will be stored. The default value uses the last analysis structure that has been created.
- *method*: the method that should be used for estimating the PSD. Allowed values are "pgram" (estimates the PSD using the FFT and optionally smooths the estimate with Daniell smoothers), "ar" (uses an AR model for the estimation) and "lomb" (Lomb-Scargle periodogram).

4.3 Frequency-Domain Analysis with RHRV

- *doPlot*: logical value, if *TRUE* (default) the periodogram is plotted using the *PlotPSD* function. Thus, if *doPlot* is *TRUE* the same parameters used in *PlotPSD* can be specified to the *CalculatePSD* function.
- Depending on the selected method, additional parameters can be specified.

As previously mentioned, the *PlotPSD* function plots the PSD estimate of the RR time series. This function highlights the different frequency bands by using colors. Thus, the *PlotPSD* function accepts as input parameters:

- *indexFreqAnalysis*: indicates which spectral analysis should be plotted. The default value uses the last analysis structure that has been created.
- The values of the band boundaries: *ULFmin*, *ULFmax*, *VLFmin*, *VLFmax*, *LFmin*, *LFmax*, *HFmin*, and *HFmax*. If no boundaries are specified, the default values are ULF = [0,0.03] Hz, VLF = [0.03,0.05] Hz, LF = [0.05,0.15] Hz, and HF = [0.15,0.4] Hz. If some band limit takes the *NULL* value, the corresponding band is not plotted.
- Usual R plotting parameters such as *log*, *xlab*, *ylab*, and *main*.
- AddLegend: if *TRUE*, a simple legend with the band boundaries is added to the plot.

4.3.1.1 Lomb-Scargle Periodogram

Listing 4.3.3 estimates and plots the PSD using the Lomb-Scargle periodogram. The plot produced by Listing 4.3.3 is shown in Fig. 4.3. Note the prominent peak centered at 0.25 Hz (plots are in logarithmic scale), which is the breathing frequency.

R listing 4.3.3

```
hrv.data <- CreateHRVData()
hrv.data <- LoadBeatWFDB(hrv.data, RecordName = "M2",
                        RecordPath = "sampleData",
                        annotator = "qrs")
hrv.data <- BuildNIHR(hrv.data)
hrv.data <- FilterNIHR(hrv.data)
hrv.data <- SetVerbose(hrv.data,TRUE)
hrv.data <- CreateFreqAnalysis(hrv.data)

## Creating frequency analysis
## Data has now 1 frequency analysis

hrv.data <- CalculatePSD(hrv.data, indexFreqAnalysis = 1,
                        method = "lomb", doPlot = FALSE)

## Calculating Periodogram using Lomb periodogram

PlotPSD(hrv.data, indexFreqAnalysis = 1)
```

After running this code, the results of the spectral analysis have been stored in the *periodogram* field of the spectral analysis data structure. Depending on the method

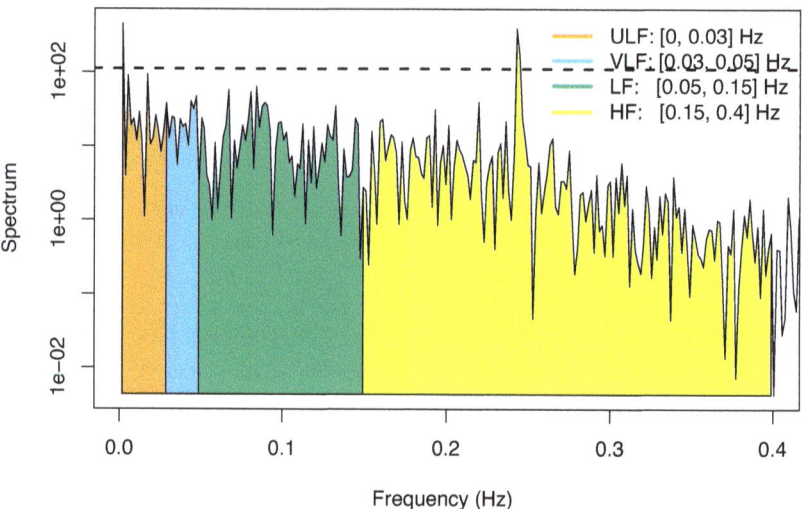

Fig. 4.3 PSD estimate obtained with the Lomb-Scargle periodogram

used for performing the estimation, the *periodogram* field will contain different slots. However, all methods share the following common fields:

- *freq*: vector of frequencies at which the spectral density is estimated
- *spec*: spectral density estimation
- *method*: method used to calculate the periodogram

Thus, *hrv.data$FreqAnalysis[[1]]$spectrogram$spec* (see Fig. 2.2) is an array containing the periodogram; the subscript *[[1]]* is due to the use of a list to store the (possibly) multiple spectral analysis data structures. Since this is the first frequency analysis, the index *1* is used.

In Listing 4.3.3, the *PlotPSD* is using default values for all the parameters that it uses. We may change basic plotting parameters (e.g., *main* and *log* parameters), colors, or the default frequency bands shown in the plot, as shown in Listing 4.3.4. It must be noted that the frequency bands do not change the PSD estimation, but they only change the representation of this estimation (see Fig. 4.4).

R listing 4.3.4

```
PlotPSD(hrv.data, indexFreqAnalysis = 1,
        usePalette = c("black", "darkorchid3",
                       "darkolivegreen4",
                       "dodgerblue4", "goldenrod1"),
        ULFmin = NULL, ULFmax = NULL, VLFmin = 0, VLFmax = 0.05,
        main = "PSD estimate using Lomb-Scargle method",
        log = "")
```

4.3 Frequency-Domain Analysis with RHRV

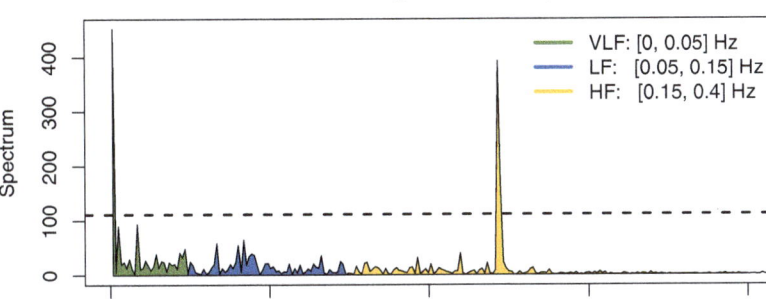

Fig. 4.4 Non-default options with the *PlotPSD* function

4.3.1.2 AR-Based Periodogram

When the *CalculatePSD* function is used for performing AR-based PSD estimation (*method="ar"*), the following parameters can be specified:

- *indexFreqAnalysis* and *doPLot*: with the same meaning as in Sect. 4.3.1.1.
- *n.freq*: the number of frequency points at which the estimation is performed (default is 500 points).
- *order*: the order of the AR model to be fitted. If omitted, a suitable order is chosen using the Akaike information criterion (AIC) [27].

Listing 4.3.5 illustrates how to calculate the periodogram using AR modeling prior to the PSD estimation. Since no order has been specified, *CalculatePSD* automatically selects a suitable order of the model. In this case, an AR(24) model has been selected, as shown in Fig. 4.5. Again, the coupling between respiration and heart rhythm produces a clear peak centered at 0.25 Hz. However, the AR spectrum is smoother than the Lomb-Scargle periodogram.

It should be noted that the AR-based PSD estimation, unlike the Lomb-Scargle periodogram, requires equally sampled time series. To derive this series from the original RR series, Listing 4.3.5 uses the function *InterpolateNIHR* (see Sect. 2.2) with a sampling frequency of 4 Hz (default value), which is sufficient to avoid spurious effects in the spectrum in most RR recordings. Note that Listing 4.3.3 did not use interpolation because the Lomb-Scargle periodogram does not require equally sampled data.

R listing 4.3.5

```
hrv.data <- CreateHRVData()
hrv.data <- LoadBeatWFDB(hrv.data, RecordName = "M2",
                        RecordPath = "sampleData",
                        annotator = "qrs")
hrv.data <- BuildNIHR(hrv.data)
```

```
hrv.data <- FilterNIHR(hrv.data)
# AR methods do need interpolated data!!
hrv.data <- InterpolateNIHR(hrv.data)
hrv.data <- SetVerbose(hrv.data, TRUE)
hrv.data <- CreateFreqAnalysis(hrv.data)

## Creating frequency analysis
## Data has now 1 frequency analysis
hrv.data <- CalculatePSD(hrv.data, indexFreqAnalysis = 1,
                    method = "ar", doPlot = TRUE, log = "")

## Calculating Periodogram using AR modelling
```

In Listing 4.3.5, *CalculatePSD* is using default values for all parameters for which it is possible. In Listing 4.3.6, an AR model of order 16 is used, as suggested by Boardman et al. [19]. The periodogram obtained using the resulting model is evaluated at *n.freq = 1000* data points (see Fig. 4.6). It should be noted the huge effect that the *order* parameter has in the estimation. The spectrum showed in Fig. 4.6 has more energy in the VLF band than the spectrum showed in Fig. 4.5. The energy of the 0.25 Hz harmonic also differs between the estimates. Indeed, this variability in the AR spectrum can be misleading. The general advice is using AR modeling when we are certain that the process that is generating the data is a stationary AR model. On the other hand, when we are uncertain about the nature of the process generating the data, a nonparametric procedure should be preferred.

R listing 4.3.6

```
# ... load, filter and interpolated data
hrv.data <- CalculatePSD(hrv.data, indexFreqAnalysis = 1,
                    method = "ar", order = 16, n.freq = 1000,
                    doPlot = TRUE, log = "")

## Calculating Periodogram using AR modelling
```

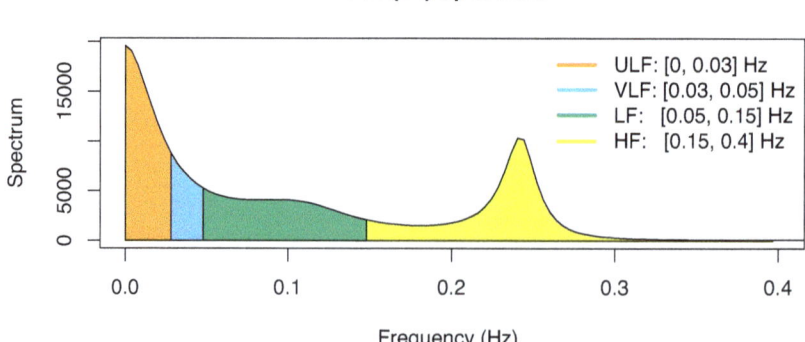

Fig. 4.5 AR-based PSD estimation. Note the smooth periodogram that is obtained with this method

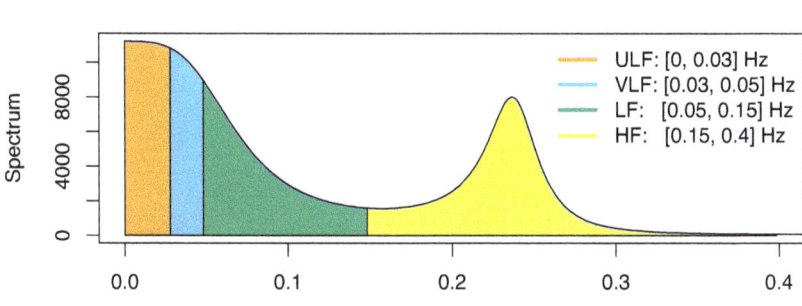

Fig. 4.6 Periodogram obtained using a fixed AR model of order 16

4.3.1.3 DFT-Based Periodogram

The most widely used methods for PSD estimation are based on the DFT. To calculate the periodogram using a FFT (and optionally smoothing) the *method* parameter must take the value *"pgram"* within the *CalculatePSD* function. In this case, the *CalculatePSD* function shares most of its parameters with the *spec.pgram* function (*stats* package). The *CalculatePSD* parameters are the following:

- *indexFreqAnalysis* and *doPLot*: with the same meaning as in Sect. 4.3.1.1.
- *spans*: vector of odd integers giving the widths of the modified Daniell smoothers to be used to smooth the periodogram. By default, no smoothing is applied.
- *kernel*: alternatively, a kernel smoother of class *tskernel*.
- *taper*: specifies the proportion of data to taper. By default, the tap is applied to 10% of the data at the beginning and at the end of the series.
- *pad*: proportion of data to pad. Zeros are added to the end of the series to increase its length by the proportion *pad*.
- *fast*: if *TRUE*, the RR series is padded to achieve a highly composite length.
- *demean*: if *TRUE*, subtract the mean of the series. Default is *FALSE*.
- *detrend*: if *TRUE*, the linear trend of the time series is removed. Default is *TRUE*.

Listing 4.3.7 estimates the periodogram by using the default values when *method= "pgram"*. Hence, the time series is detrended, taper is applied to 10% of data, and no smoothing is used (that is the reason why the resulting plot, shown in Fig. 4.7, has as title "Raw periodogram"). We can obtain smoother estimates by using longer Daniell kernels. The Daniell kernel creates a smoothed value at time t by averaging all values between times $t - m$ and $t + m$ (inclusive). We can specify the length $(2m + 1)$ of the desired Daniell kernel with the *spans* parameter. It is also possible to specify several passes of different smoothers by using a vector in the *spans* definition as shown in Listing 4.3.8. Figure 4.8 shows the PSD estimates

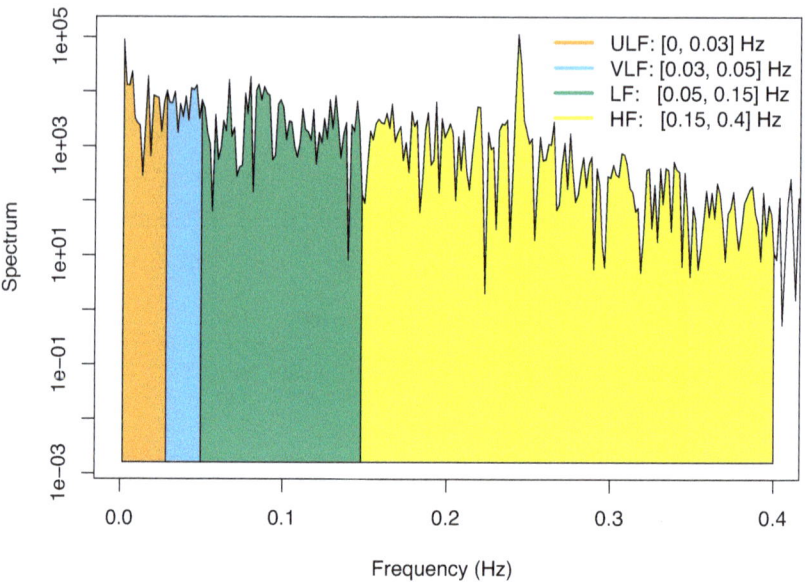

Fig. 4.7 Periodogram obtained using the DFT (no smoothing is applied)

generated by Listing 4.3.8, which are smoother than those obtained with the DFT in Fig. 4.7.

R listing 4.3.7

```
# ... load, filter and interpolated data
hrv.data <- CalculatePSD(hrv.data, indexFreqAnalysis = 1,
                         method = "pgram")

## Calculating Periodogram using DFT + Daniell smoothers
```

R listing 4.3.8

```
# ... load, filter and interpolated data
hrv.data <- CalculatePSD(hrv.data, indexFreqAnalysis = 1,
                         method = "pgram", spans = c(9, 9),
                         ULFmin = NULL)

## Calculating Periodogram using DFT + Daniell smoothers
```

Although the signal being analyzed is too short for obtaining a reliable ULF estimate, we will explore the huge impact that detrending the series has in this band for illustrative purposes (see Fig. 4.9). In Listing 4.3.9, no detrending is performed

4.3 Frequency-Domain Analysis with RHRV

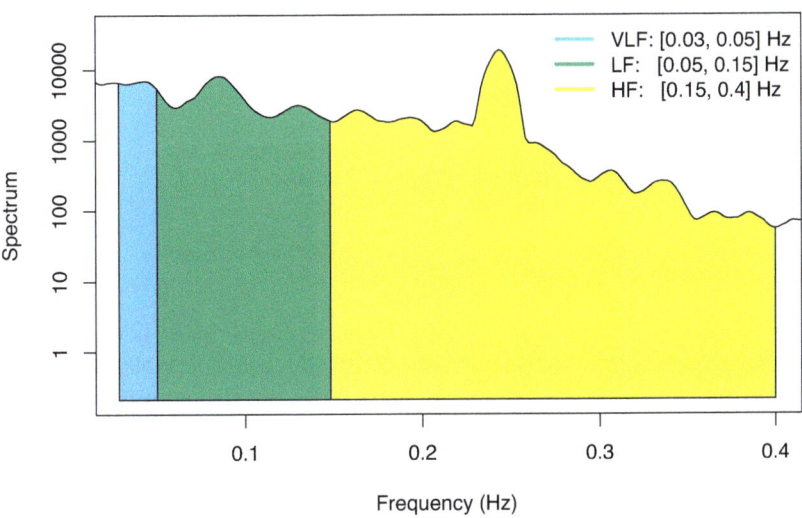

Fig. 4.8 Periodogram obtained using the DFT and two passes of a Daniell smoother of length 9

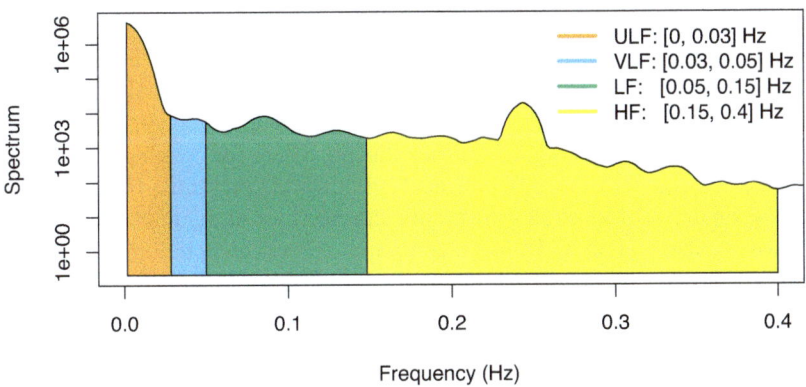

Fig. 4.9 Illustration of the effect of detrending in the PSD estimation. Compare this figure with Fig. 4.8. Since the time series has not been detrended, the energy in the ULF band dominates the periodogram

and thus, the energy in the ULF band is greater than the estimates showed in Figs. 4.7 and 4.8. Most authors recommend detrending the RR series before performing spectral analysis to avoid spurious effects in the ULF band [22].

R listing 4.3.9

```
# ... load, filter and interpolated data
hrv.data <- CalculatePSD(hrv.data, indexFreqAnalysis = 1,
                        method = "pgram", spans = c(9, 9),
                        detrend = FALSE)

## Calculating Periodogram using DFT + Daniell smoothers
```

Additional parameters or even different kernels can be used when *method="pgram"*, but for the sake of brevity, we will not explain them here. The interested reader is referred to [18] and the documentation of the *spec.pgram* function.

4.3.1.4 Creating Several Frequency Analyses

It is possible to store in the same *hrv.data* structure multiple spectral analyses. There are several reasons why this may be useful. We may be interested in performing the analyses with different parameters (e.g., different kernels or AR models). We may also be interested in using different techniques (Lomb-Scargle, AR models, or the FT) and compare the results.

We can call the function *CreateFreqAnalysis* as many times as necessary to create a data structure to store the results of each frequency analysis, and then we shall use the parameter *indexFreqAnalysis* to indicate which spectral analysis we are working with. Listing 4.3.10 uses this procedure for creating a "multiplot" in which we compare the different analysis methods presented in the previous sections. Figure 4.10 illustrates the main characteristics of the three methods. The DFT-based periodogram looks like a compromise between both the Lomb-Scargle periodogram and the AR periodogram. The Lomb-Scargle periodogram provides clear evidence of the harmonics present in the signal, whereas its estimations in the lower frequencies are usually smaller than in the rest of the methods, since it avoids the low-pass effect introduced by the resampling [22]. Finally, The AR method produces the smoothest spectrum, although as we have already argued, the spectrum can be misleading [28].

R listing 4.3.10

```
hrv.data <- CreateHRVData()
hrv.data <- LoadBeatWFDB(hrv.data, RecordName = "M2",
                        RecordPath = "sampleData",
                        annotator = "qrs")
hrv.data <- BuildNIHR(hrv.data)
hrv.data <- FilterNIHR(hrv.data)
```

4.3 Frequency-Domain Analysis with RHRV

```r
hrv.data <- SetVerbose(hrv.data, TRUE)
# Lomb-Scargle does not require interpolation ...
hrv.data <- CreateFreqAnalysis(hrv.data)

##  Creating frequency analysis
##  Data has now 1 frequency analysis

hrv.data <- CalculatePSD(hrv.data, indexFreqAnalysis = 1,
                        method = "lomb", doPlot = FALSE)

##  Calculating Periodogram using Lomb periodogram

# ... but the others methods need equally sampled data
hrv.data <- InterpolateNIHR(hrv.data)

##  Interpolating instantaneous heart rate
##  Frequency: 4 Hz
##  Number of beats: 662
##  Number of points: 2387

hrv.data <- CreateFreqAnalysis(hrv.data)

##  Creating frequency analysis
##  Data has now 2 frequency analysis

hrv.data <- CalculatePSD(hrv.data, indexFreqAnalysis = 2,
                        method = "ar", order = 16,
                        doPlot = FALSE)

##  Calculating Periodogram using AR modelling

hrv.data <- CreateFreqAnalysis(hrv.data)

##  Creating frequency analysis
##  Data has now 3 frequency analysis

hrv.data <- CalculatePSD(hrv.data, indexFreqAnalysis = 3,
                        method = "pgram",
                        doPlot = FALSE)

##  Calculating Periodogram using DFT + Daniell smoothers

hrv.data <- CreateFreqAnalysis(hrv.data)

##  Creating frequency analysis
##  Data has now 4 frequency analysis

hrv.data <- CalculatePSD(hrv.data, indexFreqAnalysis = 4,
                        method = "pgram",
                        spans = 9,
                        doPlot = FALSE)

##  Calculating Periodogram using DFT + Daniell smoothers

# Plot the results
use.ylim = c(1, 1.5e+5)
layout(matrix(1:4, 2, 2, byrow = TRUE))
PlotPSD(hrv.data, 1, addLegend = FALSE,
        addSigLevel = FALSE, ylim = use.ylim)
PlotPSD(hrv.data, 2, addLegend = FALSE,
        ylim = use.ylim)
```

```
PlotPSD(hrv.data, 3, addLegend = FALSE,
        ylim = use.ylim)
PlotPSD(hrv.data, 4, addLegend = FALSE,
        ylim = use.ylim)
```

4.3.2 Frequency Analysis of Nonstationary Signals

In this section, a recording that belongs to a patient admitted to an ICU who suffered from paraplegia and hypertension (systolic blood pressure above 200 mmHg) will be used. About 30 min after starting the recording, he is administered prostaglandin E1 (a vasodilator). His systolic blood pressure fell quickly to 100 mmHg, and then it increased slowly during almost an hour up to approximately 150 mmHg. The heart rate time series is shown in Fig. 4.11.

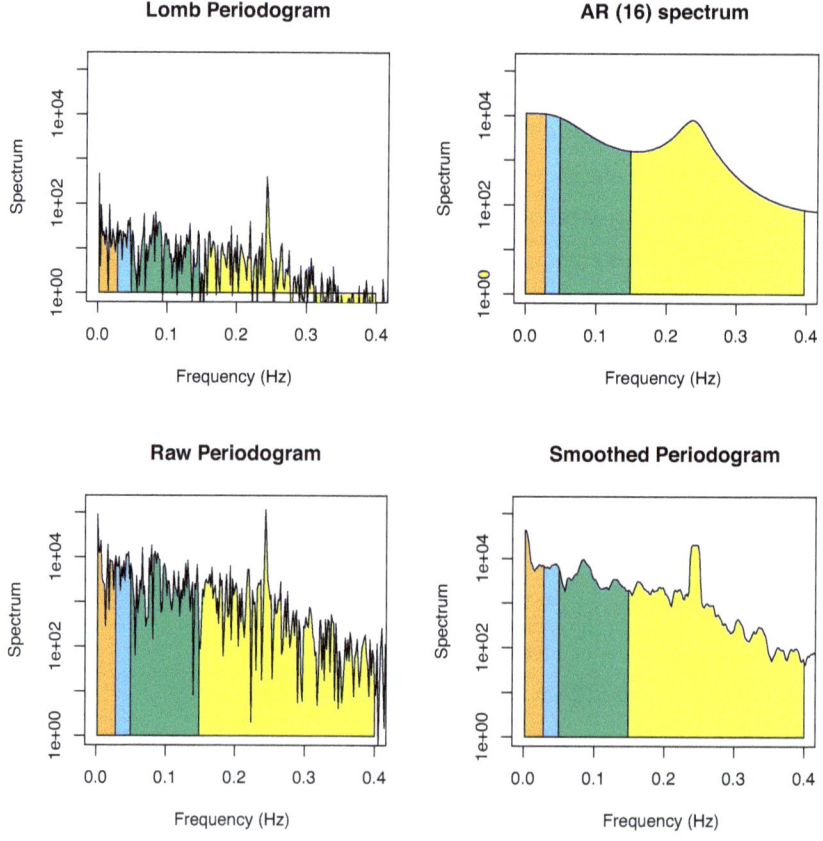

Fig. 4.10 Comparison of the different PSD estimation methods

4.3 Frequency-Domain Analysis with RHRV

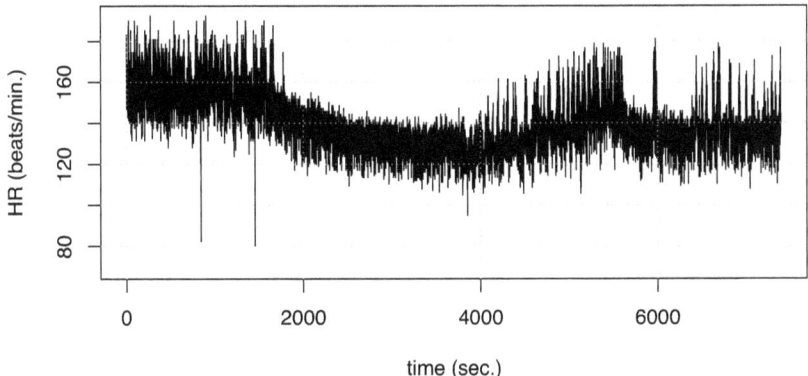

Fig. 4.11 Heart rate series used for illustrative purposes in this chapter. Note the drop in heart rate due to the prostaglandin E1 administration

This recording, which will be referred to as "example.beats," can be freely downloaded from [29] (under the *Documentation* section). Additionally, the data from this file has been included in RHRV. The user can access this data executing the code shown in Listing 4.3.11.

R listing 4.3.11

```
# hrv.data structure containing the heart beats
data("HRVData")
# HRVData structure storing the results of processing
# the heart beats: the beats have been filtered,
# interpolated, ...
data("HRVProcessedData")
```

The most important function to perform nonstationary spectral HRV analysis is the *CalculatePowerBand* function. The *CalculatePowerBand* function computes the spectrogram of the RR series in the ULF, VLF, LF, and HF frequency bands using STFT or wavelets. Boundaries of the bands may be chosen by the user by specifying the values of the function parameters *ULFmin, ULFmax, VLFmin, VLFmax, LFmin, LFmax, HFmin,* and *HFmax*. If no boundaries are specified, the default values are ULF = [0,0.03] Hz, VLF = [0.03,0.05] Hz, LF = [0.05,0.15] Hz, and HF = [0.15,0.4] Hz.

Internally the *CalculatePowerBand* function computes the spectrogram by using either STFT or wavelets. The type of analysis can be selected by the user by specifying the *type* parameter. The possible values of this parameter are *"fourier"* or *"wavelet"*. Although the analysis based on wavelets is theoretically superior and more suitable for nonstationary signals, to maintain backward compatibility in RHRV, the default value for this parameter is *"fourier"*. In the following subsections, we will see the parameters that can be specified when performing the analysis using Fourier and wavelets.

4.3.2.1 Frequency-Domain Analysis Using Fourier

When *CalculatePowerBand* uses the STFT to compute the spectrogram, the user may specify the following parameters:

- *indexFreqAnalysis*: specifies the data structure for storing the results of the spectral analysis. The default value uses the last analysis structure that has been created.
- The values of the band boundaries: *ULFmin*, *ULFmax*, *VLFmin*, *VLFmax*, *LFmin*, *LFmax*, *HFmin*, and *HFmax*.
- *size*: the size of the window used for calculating the spectrogram measured in seconds. Internally, the RHRV package uses a Hamming window when computing the STFT.
- *shift*: the displacement of window for calculating the spectrogram measured in seconds. If *shift* < *size*, there will be overlap between the windows.
- *sizesp*: the number of points for calculating each window of the STFT. It must be greater than or equal to the number of samples of the window size. It is highly recommended to select *sizesp* so that it is a highly composite number. If the user does not specify the value of *sizesp*, RHRV uses the shortest length that is a power of 2 and that is larger than the window size.

Listing 4.3.12 performs a frequency analysis in the typical HRV spectral bands, based on the STFT. It uses a 600 s window size and a 30 s displacement (these are typical values when performing HRV spectral analysis). As usual, before performing any analysis of the RR intervals, we must filter out outliers (*FilterNIHR*). As in Sects. 4.3.1.2 and 4.3.1.3, we need an equally spaced time series and hence the use of the *InterpolateNIHR* function.

R listing 4.3.12

```
hrv.data <- CreateHRVData()
hrv.data <- SetVerbose(hrv.data, FALSE)
hrv.data <- LoadBeatAscii(hrv.data, "sampleData/example.beats")
hrv.data <- BuildNIHR(hrv.data)
hrv.data <- FilterNIHR(hrv.data)
hrv.data <- InterpolateNIHR(hrv.data)
hrv.data <- CreateFreqAnalysis(hrv.data)
hrv.data <- SetVerbose(hrv.data, TRUE)
hrv.data <- CalculatePowerBand(hrv.data, size = 600,
                               shift = 30)

##  Calculating power per band
##  Using Fourier analysis
##  Windowing signal... 227 windows
##  Power per band calculated
```

After running this code, the results of the spectral analysis have been stored in the fields of the spectral analysis data structure. For example, *hrv.data$FreqAnalysis[[1]]$HRV* (see Fig. 2.2) is an array containing the total

4.3 Frequency-Domain Analysis with RHRV

spectral power for each of the windows used in the STFT; the subscript *[[1]]* is due to the use of a list to store the (possibly) multiple spectral analysis data structures. The fields *$ULF*, *$VLF*, *$LF*, and *$HF* of the data structure $FreqAnalysis[[1]] are arrays containing the power in the bands ULF, VLF, LF, and HF, respectively. The field *$LFHF* is an array containing the ratio of the power in LF and in HF. Each of these arrays contains one value for each shift of the window used in the STFT.

Given that the RR series is not stationary, it is often interesting to analyze the temporal evolution of the spectral power in the different bands. However, if the recording analyzed is short, or if we assume approximately stationary conditions over the recording, we may be interested in working with the average values of the spectral power in each of the bands. These average values are also computed by the function *CalculatePowerBand* and stored in the fields *$ULF_mean*, *$VLF_mean*, *$LF_mean*, *HF_mean* , and *$LFHF_mean (see Fig. 2.2)*.

In Listing 4.3.12, the function *CalculatePowerBand* is using default values for all the parameters for which it is possible to use a default value. We may have specified the number of points for calculating each window of the STFT and the limits of the frequency bands. The call to the function *CalculatePowerBand* in Listing 4.3.13 is equivalent to the function call in Listing 4.3.12, where all possible parameters for utilizing the STFT have been specified (although with the same values assigned by default).

R listing 4.3.13

```
hrv.data <-
  CalculatePowerBand(hrv.data, indexFreqAnalysis= 1,
                     size = 600, shift = 30,
                     sizesp = 4096,
                     type = "fourier",
                     ULFmin = 0, ULFmax = 0.03,
                     VLFmin = 0.03, VLFmax = 0.05,
                     LFmin = 0.05, LFmax = 0.15,
                     HFmin = 0.15, HFmax = 0.4 )
```

The value of the zero-padding must be greater than the number of samples of the window. Given that the sampling frequency is $f_s = 4\,\text{Hz}$ (the default value of the *InterpolateNIHR* function), the zero-padding value must fulfill $sizesp \geq size \cdot f_s$. In this occasion, we have selected the smallest power of 2 that meets the previous condition: $sizesp = 4096 = 2^{12} > 2400 = 600 \cdot 4\ (size \cdot f_s)$, hence the $4096 = 2^{12}$ value of the *sizesp* parameter.

4.3.2.2 Plotting the Spectral Bands

RHRV provides a function for plotting the spectral power of each frequency band: *PlotPowerBand*. This function can receive as inputs the following:

- *indexFreqAnalysis*: indicates which spectral analysis will be plotted. By default, the last spectral analysis that has been carried out is plotted.
- *ymax*: the maximum value of the y-axis of the power band plot.
- *ymaxratio*: the maximum value of the y-axis in the *LF/HF* plot.

R listing 4.3.14

```
PlotPowerBand(hrv.data, ymax = 200, ymaxratio = 1.7)

## Plotting power per band
## Plotted LF/HF
## Plotted ULF
## Plotted VLF
## Plotted LF
## Plotted HF
## Power per band plotted
```

Listing 4.3.14 plots the power bands computed in Listing 4.3.12. Figure 4.12 shows the plot produced by Listing 4.3.14. We can observe a significant temporal variation in the spectral power bands throughout the 2 h of recording, especially in the ULF band. Computing only average values of the spectral bands hides this type of information. Therefore, when conducting a frequency analysis, we recommend, at least, to perform a visual inspection of the temporal evolution of the spectral bands.

RHRV supplies another function for computing the spectrogram without being restricted to the four standard bands: *CalculateSpectrogram*. This function uses the STFT to compute a matrix where rows represent time and columns represent frequency. This matrix is not stored in the *HRVData* structure, since it can be very expensive in terms of memory. In the function *CalculateSpectrogram* we can specify the same *size*, *shift*, and *sizesp* parameters used in the *CalculatePowerBand* function.

We shall compute the spectrogram of the "example.beats" recording.

R listing 4.3.15

```
hrv.data <- CreateHRVData()
hrv.data <- SetVerbose(hrv.data, FALSE)
hrv.data <- LoadBeatAscii(hrv.data, "sampleData/example.beats")
hrv.data <- BuildNIHR(hrv.data)
hrv.data <- FilterNIHR(hrv.data)
hrv.data <- InterpolateNIHR (hrv.data)
hrv.data <- CreateFreqAnalysis(hrv.data)
hrv.data <- CalculatePowerBand(hrv.data, size = 600,
                               shift = 30)
spectrogram <- CalculateSpectrogram(hrv.data,
                                    size = 600,
                                    shift = 30)
```

4.3 Frequency-Domain Analysis with RHRV

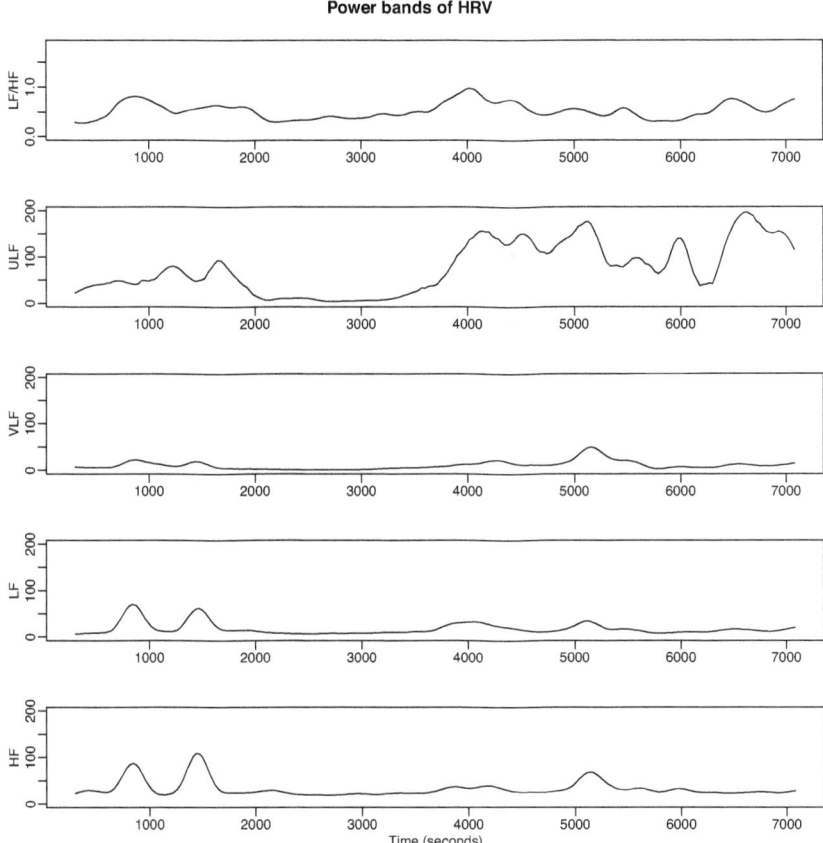

Fig. 4.12 Power bands computed using the STFT

A graphical representation of the spectrogram can be obtained with the R function *image* or with the RHRV *PlotSpectrogram* function (see Fig. 4.13). In addition to the usual *size*, *shift*, and *sizesp* parameters, the user can specify the following ones:

- *scale*: the scale used to plot spectrogram. The possible values are "*linear*" (linear axis) or "*logarithmic*" (logarithmic axis).
- *freqRange:* it specifies the minimum and maximum frequency values to represent on the y-axis. It permits obtaining a more detailed representation of a spectral band.

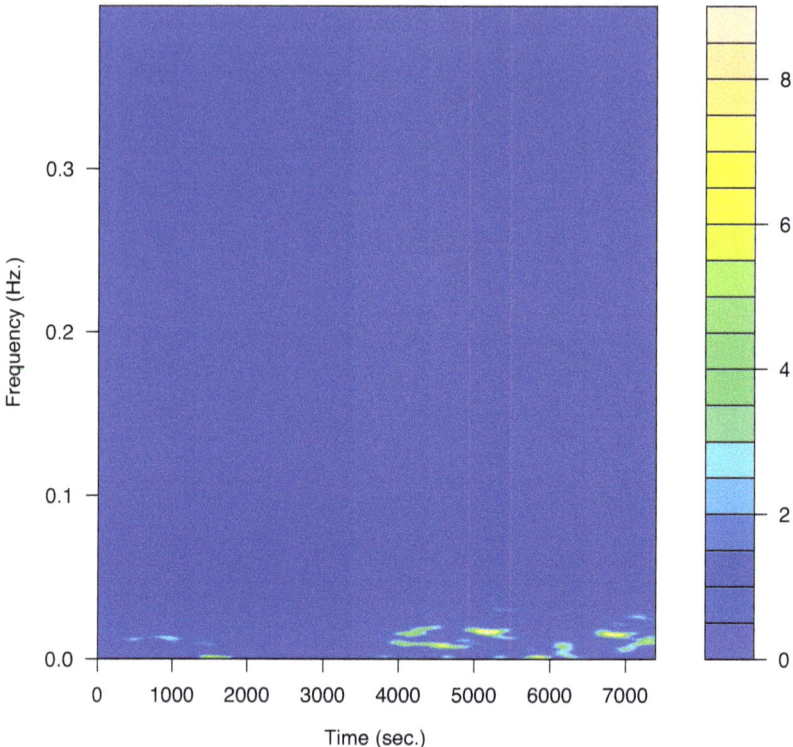

Fig. 4.13 Spectrogram of the RR time series

R listing 4.3.16

```
spectrogram <- PlotSpectrogram(HRVData = hrv.data,
                               size = 600, shift = 60,
                               scale = "logaritmic",
                               freqRange = c(0, 0.4))
## Plotting spectrogram
## Calculating spectrogram
## Window: 2400 samples (shift: 240 samples)
## Window size for calculation: 4096 samples (zero padding: 1696 samples)
## Signal size: 29592 samples
## Windowing signal with 114 windows
## Spectrogram calculated
## Spectrogram plotted
```

Figure 4.13 shows the plot produced by Listing 4.3.16. Note that most of the energy is concentrated in the low frequencies. We can also see how these low

4.3 Frequency-Domain Analysis with RHRV

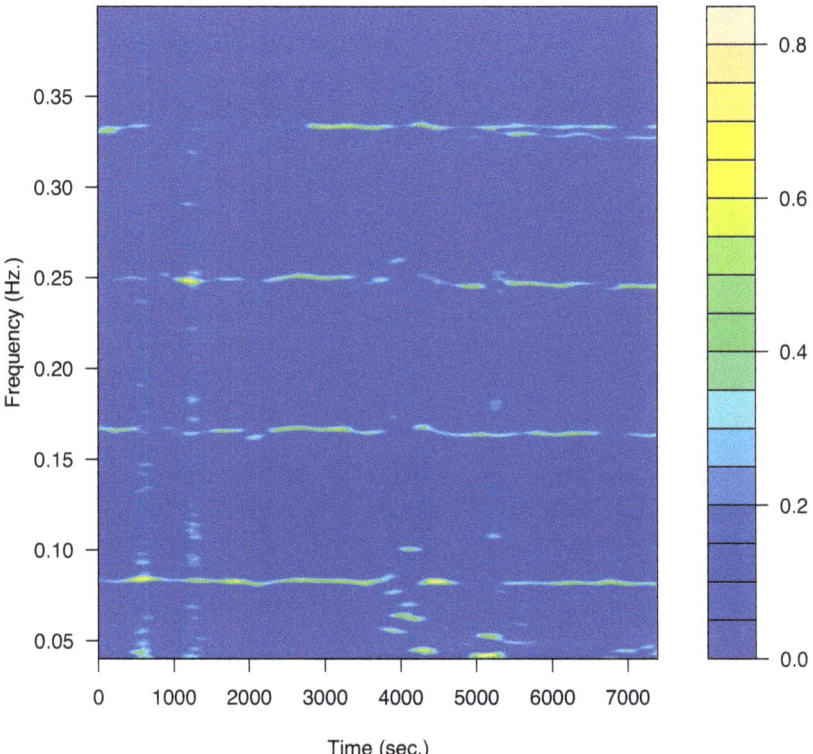

Fig. 4.14 Spectrogram ignoring the ULF band. The presence of harmonic components is apparent

frequencies, which correspond to the ULF band, decrease about a half hour into the recording, and then increase again, following the patient's blood pressure.

Since the energy is concentrated in the ULF band, the spectrogram representation may be hiding interesting information on the other bands (because its energy is negligible compared to the energy in the ULF band). Listing 4.3.17 plots the spectrogram ignoring the energy in the lowest frequencies. Figure 4.14 shows the resulting plot.

R listing 4.3.17

```
spectrogram <- PlotSpectrogram(HRVData = hrv.data,
                               size = 600, shift = 60,
                               scale = "logaritmic",
                               freqRange = c(0.04, 0.4))
```

4.3.2.3 Frequency-Domain Analysis Using Wavelets

The *CalculatePowerBand* function can compute the spectral power in the ULF, VLF, LF, and HF bands using the WT. In this case, the *type* parameter must take the value "wavelet" (the default value is "*fourier*"). When wavelets are used for the computation of the spectral power, the following parameters can be specified:

- *indexFreqAnalysis*: specifies the data structure for storing the results of the spectral analysis. By default, the results are stored in the last spectral structure that has been created.
- The values of the band boundaries: *ULFmin, ULFmax, VLFmin, VLFmax, LFmin, LFmax, HFmin*, and *HFmax*.
- *wavelet*: the mother wavelet used to calculate the spectrogram. Some of the most widely used wavelets are available: Haar ("*haar*"), extremal phase ("*d4*," "*d6*," "*d8*," and "*d16*"), the least asymmetric Daubechies ("*la8*," "*la16*," and "*la20*"), and the best localized Daubechies ("*bl14*" and "*bl20*") wavelets, among others. The default value is "*d4*."
- *bandtolerance*: maximum error allowed for the band boundaries when the wavelet-based analysis is performed (see more details in [24, 30]). It can be specified as either an absolute or a relative error. The default value is 0.01.
- *relative*: a logic value specifying which type of band tolerance error shall be used: relative (in percentage) or absolute (this is the default value).

In the *wavelet* parameter, the name of the wavelet specifies the family (the family determines the shape of the wavelet and its properties) and the length of the wavelet. For example, "*la8*" belongs to the least asymmetric family and has a length of eight samples. Shorter wavelets usually have better temporal resolution, but they will tend to fail when discriminating close frequencies. On the other hand, longer wavelets usually have worse temporal resolution, but they provide better frequency resolution. Better temporal resolution means that we can study shorter time intervals. Depending on the goal of the analysis, we can choose longer or shorter wavelets within a wavelet family.

Let $[f_l, f_u]$ be any frequency band specified by the user, and let $[f_1, f_2]$ be a frequency interval associated with some node in the maximal overlap discrete wavelet packet transform (MODWPT) tree [31]. The relative error ϵ_r of f_l over the $[f_1, f_2]$ interval is computed as

$$\epsilon_r = \left| \frac{f_l - f_1}{f_u - f_l} \right| \cdot 100\%.$$

Similarly, the relative error ϵ_r of the upper frequency f_u is

$$\epsilon_r = \left| \frac{f_u - f_2}{f_u - f_l} \right| \cdot 100\%.$$

4.3 Frequency-Domain Analysis with RHRV

The use of a relative error (*relative=TRUE*) has the advantage of avoiding large errors in the lower-frequency bands (ULF and VLF bands). The absolute value ϵ is defined as usual: $\epsilon = |f_2 - f_u|$ for the upper frequency and $\epsilon = |f_1 - f_l|$ for the lower frequency. For implementation details of how the spectral power in each band is estimated from the coefficients of the MODWPT tree, we refer the reader to [24].

We will repeat the analysis of Listing 4.3.18 using wavelets. We will use an absolute tolerance of 0.01 Hz and the least asymmetric Daubechies of width 4 ("*d4*," the default value).

R listing 4.3.18

```
hrv.data <- CreateHRVData()
hrv.data <- SetVerbose(hrv.data, FALSE)
hrv.data <- LoadBeatAscii(hrv.data, "example.beats")
hrv.data <- BuildNIHR(hrv.data)
hrv.data <- FilterNIHR(hrv.data)
hrv.data <- InterpolateNIHR (hrv.data)
hrv.data <- CreateFreqAnalysis(hrv.data)
hrv.data <- SetVerbose(hrv.data, TRUE)
hrv.data <- CalculatePowerBand(hrv.data,
                               size = 600, shift = 30)
hrv.data <- CalculatePowerBand(hrv.data,
                               type = "wavelet",
                               bandtolerance = 0.01,
                               relative = FALSE)
```

In Listing 4.3.18, the function *CalculatePowerBand* is using default values for all parameters for which it is possible to use a default value. The call to the function *CalculatePowerBand* in Listing 4.3.19 is equivalent to the function call shown in Listing 4.3.19.

R listing 4.3.19

```
hrv.data <-
  CalculatePowerBand(hrv.data, indexFreqAnalysis= 1,
                     type = "wavelet", wavelet = "d4",
                     bandtolerance = 0.01,
                     relative = FALSE,
                     ULFmin = 0, ULFmax = 0.03,
                     VLFmin = 0.03, VLFmax = 0.05,
                     LFmin = 0.05, LFmax = 0.15,
                     HFmin = 0.15, HFmax = 0.4)
```

4.3.2.4 Creating Several Frequency Analysis

As shown in Sect. 4.3.1.4, we can create multiple analyses with the *CreateFreqAnalysis* function and store them in the same *HRVData* structure. To perform this task, we shall call the function *CreateFreqAnalysis* as many times as necessary, as shown in Listing 4.3.20.

R listing 4.3.20

```
hrv.data <- CreateHRVData()
hrv.data <- LoadBeatAscii(hrv.data, "sampleData/example.beats")
hrv.data <- BuildNIHR(hrv.data)
hrv.data <- FilterNIHR(hrv.data)
hrv.data <- InterpolateNIHR(hrv.data, freqhr = 4)
hrv.data <- CreateFreqAnalysis(hrv.data)
# use freqAnalysis number 1 with Fourier analysis
hrv.data <- CalculatePowerBand(hrv.data,
                               indexFreqAnalysis = 1,
                               size = 600, shift = 30,
                               type = "fourier")
PlotPowerBand(hrv.data, indexFreqAnalysis = 1,
              ymax = 200, ymaxratio = 1.7)
# use freqAnalysis number 2 with wavelet analysis
hrv.data <- CreateFreqAnalysis(hrv.data)
hrv.data <- CalculatePowerBand(hrv.data,
                               indexFreqAnalysis = 2,
                               type = "wavelet",
                               wavelet = "la8",
                               bandtolerance = 0.01,
                               relative = FALSE)
PlotPowerBand(hrv.data, indexFreqAnalysis = 2,
              ymax = 800, ymaxratio = 50)
```

Figure 4.15a, b show the plots produced by Listing 4.3.20. They illustrate important differences between Fourier- and wavelet-based frequency analyses. The power range in both plots is not the same due to the windowing used in both techniques. Thus, we should avoid direct comparisons between the numerical results obtained with Fourier with those obtained using wavelets. Fourier's power spectrum is smoother than the wavelet's power spectrum. This is a consequence of the higher temporal resolution that the wavelet-based analysis provides. The Fourier power spectrum has a smaller number of samples than the original signal as a consequence of the use of windows. Conversely, the wavelet power spectrum has the same number of samples as the original RR time series. We could try to increase Fourier's temporal resolution by decreasing the window' size used in the analysis: the shorter window we use, the sharper spectrum we get. However, when using smaller windows, spectral resolution decreases.

4.3 Frequency-Domain Analysis with RHRV

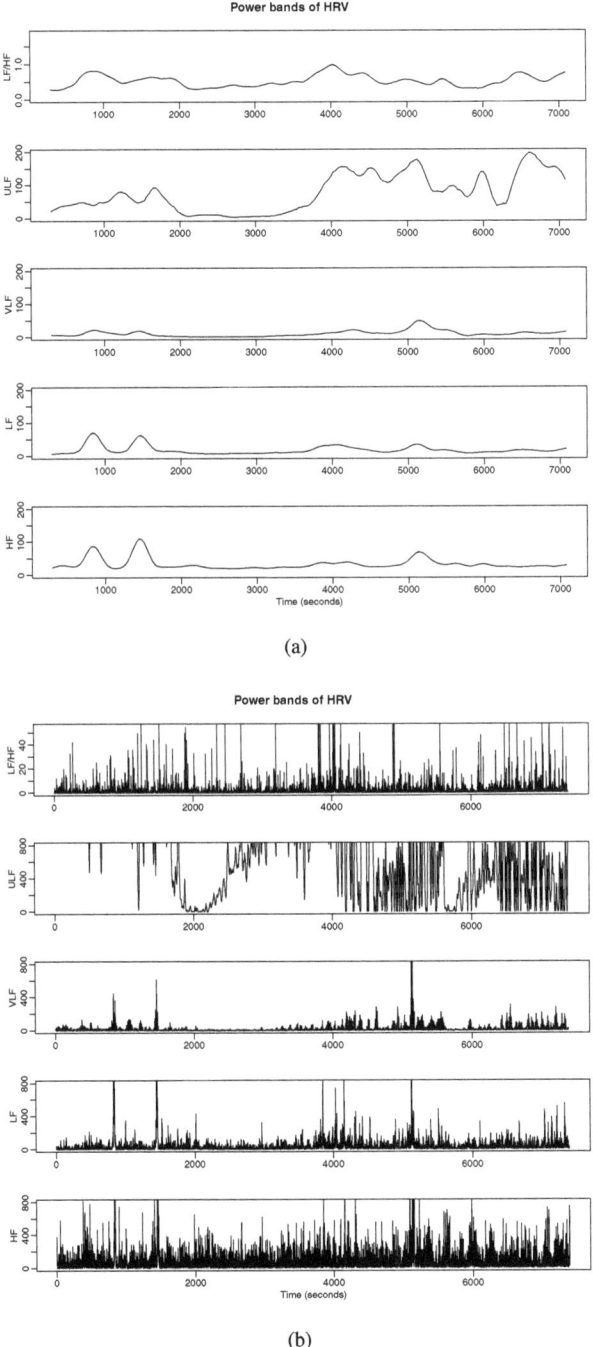

Fig. 4.15 Analysis of the RR series using both Fourier and the wavelet transform. (**a**) Fourier analysis. (**b**) Wavelet analysis

4.4 Changes in HRV Frequency-Based Statistics Under Pathological Conditions

HRV power spectrum follows the general rule that the more healthy a heart is, the more variability it has. Typically, under pathological conditions, there is often a decrease on the spectral power in all bands. This decrease tends to affect more often the HF band. It is also common that under pathological conditions, there is increased sympathetic activity either in absolute terms (and therefore it is associated with an increase in total power in the LF band) or in relative terms (and therefore it is associated with an increase in the ratio LF/HF). In the following paragraphs, we summarize some of the changes in HRV power spectrum under pathological conditions for which there is considerable evidence in the literature, although this section is far from being an exhaustive list.

After myocardial infarction, patients present reduced overall spectral power and a reduced power in each frequency band [32, 33], although there is a relative increase in the power in the LF band compared to the HF band. These patients also have an increased LF/HF pointing to a change in the sympathetic-vagal balance toward the predominance of sympathetic tone. Spectral HRV is a predictor of mortality and of arrhythmic complications in patients following acute myocardial infarction and can be used for risk stratification [11].

Patients suffering from cardiomyopathies [34] and from supraventricular arrhythmias [35] have total spectral power and power in the HF band reduced. Patients suffering from ventricular arrhythmias present a reduction in total spectral power before the onset of sustained ventricular tachycardia when compared with the onset of a nonsustained ventricular tachycardia [36]. Patients suffering from mitral valve prolapse present a decrease in power in the HF band [37].

Right after a heart transplant, patients have very reduced power in all spectral bands [38, 39]. As reinnervation progresses after the transplant, the spectral power increases, although it still remains reduced compared with healthy patients.

Patients suffering from congestive heart failure present reduced spectral power in all frequencies, especially below 0.04 Hz [40, 41]. They also present an increased LF/HF pointing to a predominance of sympathetic tone [42].

Patients at risk of suffering sudden death or cardiac arrest present an increase in power in the LF band and a decrease in power in the HF band [43, 44], pointing to an increased sympathetic activity compared to control patients, and a decreased vagal activity. Similar findings appear in patients with hypertension [45, 46].

Patients with diabetic neuropathy present a reduction of the absolute power in the bands LF and HF [47] although the ratio LF/HF is not altered.

Obstructive sleep apnea syndrome produces an increase in total spectral power and in the VLF and LF bands. The LF/HF ratio is also increased [48, 49].

References

1. S. Akselrod, D. Gordon, F.A. Ubel, D.C. Shannon, A. Berger, R.J. Cohen, Power spectrum analysis of heart rate fluctuation: a quantitative probe of beat-to-beat cardiovascular control. Science **213**(4504), 220–222 (1981). [Online]. Available: https://doi.org/10.1126/science.616604
2. S. Babaeizadeh, S. Zhou, X. Liu, W. Hu, D. Feild, E. Helfenbein, R. Gregg, J. Lindauer, A novel heart rate variability index for evaluation of left ventricular function using five-minute electrocardiogram, in *Computers in Cardiology, 2007* (IEEE, Piscataway, 2007), pp. 473–476. [Online]. Available: https://doi.org/10.1109/CIC.2007.4745525
3. R. Bailón, J. Mateo, S. Olmos, P. Serrano, J. García, A. Del Río, I. Ferreira, P. Laguna, Coronary artery disease diagnosis based on exercise electrocardiogram indexes from repolarisation, depolarisation and heart rate variability. Med. Biol. Eng. Comput. **41**(5), 561–571 (2003). [Online]. Available: https://doi.org/10.1007/BF02345319
4. A. Bickel, M. Yahalom, N. Roguin, R. Frankel, J. Breslava, S. Ivry, A. Eitan, Power spectral analysis of heart rate variability during positive pressure pneumoperitoneum. Surg. Endosc. Interventional Tech. **16**(9), 1341–1344 (2002). [Online]. Available: https://doi.org/10.1007/s00464-001-9211-6
5. A. Bickel, M. Yahalom, N. Roguin, S. Ivry, J. Breslava, R. Frankel, A. Eitan, Improving the adverse changes in cardiac autonomic nervous control during laparoscopic surgery, using an intermittent sequential pneumatic compression device. Am. J. Surg. **187**(1), 124–127 (2004). [Online]. Available: https://doi.org/10.1016/j.amjsurg.2003.02.008
6. D. Gordon, V.L. Herrera, L. McAlpine, R.J. Cohen, S. Akselrod, P. Lang, W.I. Norwood, Heart-rate spectral analysis: a noninvasive probe of cardiovascular regulation in critically ill children with heart disease. Pediatr. Cardiol. **9**(2), 69–77 (1988). [Online]. Available: https://doi.org/10.1007/BF02083703
7. Y. Arai, J.P. Saul, P. Albrecht, L.H. Hartley, L.S. Lilly, R.J. Cohen, W.S. Colucci, Modulation of cardiac autonomic activity during and immediately after exercise. Am. J. Physiol. **256**(1), H132–H141 (1989). [Online]. Available: https://doi.org/10.1152/ajpheart.1989.256.1.H132
8. M. Kuwahara, K.-i. Yayou, K. Ishii, S.-i. Hashimoto, H. Tsubone, S. Sugano, Power spectral analysis of heart rate variability as a new method for assessing autonomic activity in the rat. Electrocardiology **27**(4), 333–337 (1994). [Online]. Available: https://doi.org/10.1016/s0022-0736(05)80272-9
9. P.P. Pereira-Junior, E.A. Chaves, R.H. Costa-e Sousa, M.O. Masuda, A.C.C. de Carvalho, J.H. Nascimento, Cardiac autonomic dysfunction in rats chronically treated with anabolic steroid. Eur. J. Appl. Physiol. **96**(5), 487–494 (2006). [Online]. Available: https://doi.org/10.1007/s00421-005-0111-7
10. F. Yasuma, J.-i. Hayano, Respiratory sinus arrhythmia: why does the heartbeat synchronize with respiratory rhythm? Chest J. **125**(2), 683–690 (2004). [Online]. Available: https://doi.org/10.1378/chest.125.2.683
11. Task Force of the European Society of Cardiology and the North American Society of Pacing and Electrophysiology. Heart rate variability: standards of measurement, physiological interpretation and clinical use. Eur. Heart J. **17**, 354–381 (1996). [Online]. Available: https://doi.org/10.1093/oxfordjournals.eurheartj.a014868
12. M.L. Appel, R.D. Berger, J.P. Saul, J.M. Smith, R.J. Cohen, Beat to beat variability in cardiovascular variables: noise or music? J. Am. Coll. Cardiol. **14**(5), 1139–1148 (1989). [Online]. Available: https://doi.org/10.1016/0735-1097(89)90408-7
13. M.V. Kamath, E.L. Fallen, Power spectral analysis of heart rate variability: a noninvasive signature of cardiac autonomic function. Crit. Rev. Biomed. Eng. **21**(3), 245–311 (1992)
14. A. Malliani, M. Pagani, F. Lombardi, S. Cerutti, Cardiovascular neural regulation explored in the frequency domain. Circulation **84**(2), 482–492 (1991). [Online]. Available: https://doi.org/10.1161/01.cir.84.2.482

15. G.G. Berntson, J.T. Cacioppo, Heart rate variability: stress and psychiatric conditions. Dyn. Electrocardiography, 57–64 (2004). [Online]. Available: https://doi.org/10.1002/9780470987483.ch7
16. M. Nishikino, T. Matsunaga, K. Yasuda, T. Adachi, T. Moritani, G. Tsujimoto, K. Tsuda, N. Aoki, Genetic variation in the renin-angiotensin system and autonomic nervous system function in young healthy japanese subjects. J. Clin. Endocrinol. Metab. **91**(11), 4676–4681 (2006). [Online]. Available: https://doi.org/10.1210/jc.2006-0700
17. J.G. Proakis, D.G. Manolakis, *Introduction to Digital Signal Processing* (Prentice Hall Professional Technical Reference, Hoboken, 1988)
18. A.V. Metcalfe, P.S. Cowpertwait, *Introductory Time Series with R* (Springer, Berlin, 2009). [Online]. Available: https://doi.org/10.1007/978-0-387-88698-5
19. A. Boardman, F.S. Schlindwein, A.P. Rocha, A study on the optimum order of autoregressive models for heart rate variability. Physiol. Meas. **23**(2), 325–336 (2002). [Online]. Available: https://doi.org/10.1088/0967-3334/23/2/308
20. N.R. Lomb, Least-squares frequency analysis of unequally spaced data. Astrophys. Space Sci. **39**(2), 447–462 (1976). [Online]. Available: https://doi.org/10.1007/BF00648343
21. J.D. Scargle, Studies in astronomical time series analysis. II-statistical aspects of spectral analysis of unevenly spaced data. Astrophys. J. **263**, 835–853 (1982). [Online]. Available: https://doi.org/10.1086/160554
22. P. Laguna, G.B. Moody, R.G. Mark, Power spectral density of unevenly sampled data by least-square analysis: performance and application to heart rate signals. IEEE Trans. Biomed. Eng. **45**(6), 698–715 (1998). [Online]. Available: https://doi.org/10.1109/10.678605
23. D. Gabor, Theory of communication. Part 1: the analysis of information. Inst. Elect. Eng. Part III: Radio Commun. Eng. **93**(26), 429–441 (1946). [Online]. Available: https://doi.org/10.1049/ji-3-2.1946.0074
24. C.A. García, A. Otero, X. Vila, D.G. Márquez, A new algorithm for wavelet-based heart rate variability analysis. Biomed. Signal Proces. Control **8**(6), 542–550 (2013). [Online]. Available: https://doi.org/10.1016/j.bspc.2013.05.006
25. Exaggerated heart rate oscillations during two meditation techniques. Last accessed 10 Apr 2024. [Online]. Available: http://www.physionet.org/physiobank/database/meditation/data/
26. C.-K. Peng, J.E. Mietus, Y. Liu, G. Khalsa, P.S. Douglas, H. Benson, A.L. Goldberger, Exaggerated heart rate oscillations during two meditation techniques. Int. J. Cardiol. **70**(2), 101–107 (1999). [Online]. Available: https://doi.org/10.1016/s0167-5273(99)00066-2
27. H. Akaike, Information theory and an extension of the maximum likelihood principle, in *Selected Papers of Hirotugu Akaike* (Springer, Berlin, 1998), pp. 199–213. [Online]. Available: https://doi.org/10.1007/978-1-4612-1694-0_15
28. D. Thomson, Time series analysis of holocene climate data. Philos. Trans. R. Soc. London, Ser. A **330**(1615), 601–616 (1990). [Online]. Available: https://doi.org/10.1098/rsta.1990.0041
29. C.A. García, A. Otero, M. Lado, A. Méndez, L. Rodríguez-Linares, X. Vila, The RHRV project homepage. Last accessed 10 Apr 2024. [Online]. Available: http://rhrv.r-forge.r-project.org/
30. C.A. García, A. Otero, X.A. Vila, M.J. Lado, An open source tool for heart rate variability wavelet-based spectral analysis, in *BIOSIGNALS 2012 - Proceedings of the International Conference on Bio-inspired Systems and Signal Processing, Vilamoura, Algarve, 1–4 February, 2012*, vol. 2012 (2012), pp. 206–211. [Online]. Available: https://doi.org/10.5220/0003725002060211
31. D.B. Percival, A.T. Walden, *Wavelet Methods for Time Series Analysis* (Cambridge University Press, Cambridge, 2006). [Online]. Available: https://doi.org/10.1017/CBO9780511841040
32. R.M. Carney, J.A. Blumenthal, K.E. Freedland, P.K. Stein, W.B. Howells, L.F. Berkman, L.L. Watkins, S.M. Czajkowski, J. Hayano, P.P. Domitrovich, et al., Low heart rate variability and the effect of depression on post–myocardial infarction mortality. Arch. Intern. Med. **165**(13), 1486–1491 (2005). [Online]. Available: https://doi.org/10.1001/archinte.165.13.1486
33. R.E. Kleiger, J.P. Miller, J.T. Bigger Jr, A.J. Moss, Decreased heart rate variability and its association with increased mortality after acute myocardial infarction. Am. Cardiol. **59**(4), 256–262 (1987). [Online]. Available: https://doi.org/10.1016/0002-9149(87)90795-8

34. P.J. Counihan, L. Fei, Y. Bashir, T.G. Farrell, G.A. Haywood, W.J. McKenna, Assessment of heart rate variability in hypertrophic cardiomyopathy. Association with clinical and prognostic features. Circulation **88**(4), 1682–1690 (1993). [Online]. Available: https://doi.org/10.1161/01.CIR.88.4.1682
35. D.Z. Kocovic, T. Harada, J.B. Shea, D. Soroff, P.L. Friedman, Alterations of heart rate and of heart rate variability after radiofrequency catheter ablation of supraventricular tachycardia. Delineation of parasympathetic pathways in the human heart. Circulation **88**(4), 1671–1681 (1993). [Online]. Available: https://doi.org/10.1161/01.cir.88.4.1671
36. H. Huikuri, J. Valkama, K. Airaksinen, T. Seppänen, K. Kessler, J. Takkunen, R. Myerburg, Frequency domain measures of heart rate variability before the onset of nonsustained and sustained ventricular tachycardia in patients with coronary artery disease. Circulation **87**(4), 1220–1228 (1993). [Online]. Available: https://doi.org/10.1161/01.cir.87.4.1220
37. K.M. Stein, J.S. Borer, C. Hochreiter, P.M. Okin, E.M. Herrold, R.B. Devereux, P. Kligfield, Prognostic value and physiological correlates of heart rate variability in chronic severe mitral regurgitation. Circulation **88**(1), 127–135 (1993). [Online]. Available: https://doi.org/10.1161/01.CIR.88.1.127
38. U.R. Acharya, K.P. Joseph, N. Kannathal, C.M. Lim, J.S. Suri, Heart rate variability: a review. Med. Biol. Eng. Comput. **44**(12), 1031–1051 (2006). [Online]. Available: https://doi.org/10.1007/s11517-006-0119-0
39. K. Sands, M.L. Appel, L.S. Lilly, F.J. Schoen, G. Mudge, R.J. Cohen, Power spectrum analysis of heart rate variability in human cardiac transplant recipients. Circulation **79**(1), 76–82 (1989). [Online]. Available: https://doi.org/10.1161/01.CIR.79.1.76
40. P.F. Binkley, E. Nunziata, G.J. Haas, S.D. Nelson, R.J. Cody, Parasympathetic withdrawal is an integral component of autonomic imbalance in congestive heart failure: demonstration in human subjects and verification in a paced canine model of ventricular failure. J. Am. Coll. Cardiol. **18**(2), 464–472 (1991). [Online]. Available: https://doi.org/10.1016/0735-1097(91)90602-6
41. S. Guzzetti, M.T. La Rovere, G.D. Pinna, R. Maestri, E. Borroni, A. Porta, A. Mortara, A. Malliani, Different spectral components of 24 h heart rate variability are related to different modes of death in chronic heart failure. Eur. Heart J. **26**(4), 357–362 (2005). [Online]. Available: https://doi.org/10.1093/eurheartj/ehi067
42. P.F. Binkley, G.J. Haas, R.C. Starling, E. Nunziata, P.A. Hatton, C.V. Leier, R.J. Cody, Sustained augmentation of parasympathetic tone with angiotensin-converting enzyme inhibition in patients with congestive heart failure. J. Am. Coll. Cardiol. **21**(3), 655–661 (1993). [Online]. Available: https://doi.org/10.1016/0735-1097(93)90098-L
43. A. Algra, J. Tijssen, J. Roelandt, J. Pool, J. Lubsen, Heart rate variability from 24-hour electrocardiography and the 2-year risk for sudden death. Circulation **88**(1), 180–185 (1993). [Online]. Available: https://doi.org/10.1161/01.cir.88.1.180
44. C.M. Dougherty, R.L. Burr, Comparison of heart rate variability in survivors and nonsurvivors of sudden cardiac arrest. Am. Cardiol. **70**(4), 441–448 (1992). [Online]. Available: https://doi.org/10.1016/0002-9149(92)91187-9
45. S. Guzzetti, S. Dassi, M. Pecis, R. Casat, A.M. Masu, P. Longoni, M. Tinelli, S. Cerutti, M. Pagani, A. Malliani, Altered pattern of circadian neural control of heart period in mild hypertension. J. Hypertens. **9**(9), 831–838 (1991). [Online]. Available: https://doi.org/10.1097/00004872-199109000-00010
46. W. Langewitz, H. Rüddel, H. Schächinger, Reduced parasympathetic cardiac control in patients with hypertension at rest and under mental stress. Am. Heart J. **127**(1), 122–128 (1994). [Online]. Available: https://doi.org/10.1016/0002-8703(94)90517-7
47. M. Pagkalos, N. Koutlianos, E. Kouidi, E. Pagkalos, K. Mandroukas, A. Deligiannis, Heart rate variability modifications following exercise training in type 2 diabetic patients with definite cardiac autonomic neuropathy. Br. Sports Med. **42**(1), 47–54 (2008). [Online]. Available: https://doi.org/10.1136/bjsm.2007.035303

48. M. Kesek, K.A. Franklin, C. Sahlin, E. Lindberg, Heart rate variability during sleep and sleep apnoea in a population based study of 387 women. Clin. Physiol. Funct. Imaging **29**(4), 309–315 (2009). [Online]. Available: https://doi.org/10.1111/j.1475-097X.2009.00873.x
49. D.-H. Park, C.-J. Shin, S.-C. Hong, J. Yu, S.-H. Ryu, E.-J. Kim, H.-B. Shin, B.-H. Shin, Correlation between the severity of obstructive sleep apnea and heart rate variability indices. Korean Med. Sci. **23**(2), 226–231 (2008). [Online]. Available: https://doi.org/10.3346/jkms.2008.23.2.226

Chapter 5
Nonlinear and Fractal Analysis

Abstract There are many complex systems in nature that can be explained by nonlinear interactions of its main components. The heart rate regulation is one of the most complex systems in the human body. Heart rhythm is innervated by both the parasympathetic and sympathetic branches of the ANS. At the same time, the autonomic nervous system is influenced by humoral effects, hemodynamic variables, respiratory rhythm, and stroke volume, among others. Furthermore, there exist feedback loops among these mechanisms influencing each other in a nonlinear way. A consequence of these nonlinear interactions is that heart rate modulation cannot be fully understood by studying its components in isolation. However, the study of the heart rhythm modulation as a whole is a formidable task. A more common approach in the literature, more modest than the full comprehension of the heart rate regulation system, is trying to quantify the complexity of heart rate using nonlinear statistics derived from the chaos theory or from fractal processes. In this chapter, we summarize the most widely used statistics based on nonlinear and fractal dynamics, and we show how to compute them with RHRV.

5.1 An Overview of Nonlinear Dynamics

Most nonlinear time-series analysis techniques used in the HRV literature are motivated by the theory of dynamical systems. A dynamical system is represented by a set of n state variables (that can be represented as coordinates in $\mathbb{R}^n$) that fully describe the state of the system at any time and a dynamical rule that specifies the future values of all the state variables. The collection of coordinates is usually referred to as the *state space* or *phase space* if these coordinates form a smooth manifold. There can be multiple spaces of states representing a dynamical system; i.e., the system representation is not unique. Dynamical systems can be deterministic (the future state is fully determined by the dynamical rule) or stochastic/random (there is a probability distribution defined over the future states).

In the context of HRV analysis, many successful approaches assume that the RR series is a deterministic nonlinear process (with possibly a small linear random perturbation). In this chapter, when employing techniques rooted in deterministic nonlinear dynamics, we assume that, following an initial transient period, the RR series converges to a subset in the phase space called *attractor*. Furthermore, we will focus on discrete deterministic dynamical systems, given that the major approach in HRV literature focuses on studying how the current RR interval, RR_i, determines the next one, RR_{i+1}.

Nonlinear stochastic systems can be viewed as a combination of deterministic systems and stochastic models. There is a low use of nonlinear stochastic models in HRV due to the success in the application of a particular type of stochastic processes to the RR series: fractal models. In Sect. 5.3, we give a brief overview of the main theoretical concepts of fractal models. The reader interested in the application of nonlinear stochastic processes to time-series analysis is referred to [1].

5.2 Chaotic Nonlinear Statistics

5.2.1 Nonlinearity Tests

Before studying a system using nonlinear analysis techniques, we must check that the data shows some degree of nonlinearity.

The main idea for testing for nonlinearity is the following. A statistic μ quantifying some nonlinear feature of the data is computed. Does the value found for μ (μ_{data}) suggest nonlinearity or can we explain this value with a simpler hypothesis? (e.g., a comparable linear model (this is usually the *null hypothesis* to test against)). If we know the distribution of μ for a linear process, we can test if μ_{data} gives enough evidence for assuming a nonlinear model. However, we do not usually know the distribution of μ under the null hypothesis. The preferred approach in the literature is estimating this distribution by using an ensemble of *surrogate data* [2, 3].

This ensemble is generated in such a way that each realization shares the linear properties of the original time series. For a two-sided test and a level of significance α, $2K/\alpha - 1$ surrogates are usually generated (K is an integer, $K \geq 1$, controlling the final number of surrogates). Under the null hypothesis, there are $2K/\alpha$ samples following the same distribution. Thus, each of this samples has a probability of $\alpha/2$ of yielding one of the smallest K values of μ and a probability of $\alpha/2$ of yielding one of the largest values of μ. The null hypothesis is then rejected when the observed time series yields either one of the K smallest or largest values. For a one-sided test, only $K/\alpha - 1$ surrogates are needed, and we reject the null hypothesis when the time series yields one of the K smallest/largest value of μ (depending on the particular hypothesis).

5.2.2 Phase Space Reconstruction

When observing a highly complex system (such as the HR modulation system), typically we are only able to measure a few of the dynamical variables governing the evolution of the system. In the case of HRV analysis, we usually only measure the RR intervals instead of the (unknown) set of state variables. We need a mechanism to derive the state space from the RR intervals. This problem receives the name of *phase space reconstruction*, and it is formally solved by using the *embedding theorem* [4, 5]. This theorem states that a phase space can be reconstructed from the single measured discrete series x_i by using the vectors:

$$\mathbf{x}_i = \left[x_i, x_{i+\tau}, \ldots, x_{i+(m-1)\cdot\tau} \right], \tag{5.1}$$

where m and τ will be referred to as the *embedding dimension* and the *time lag parameter*. The vectors $\mathbf{x}_i$ constructed using Eq. 5.1 will be denominated *Takens' vectors*.

Great effort has been made to clarify under which circumstances Eq. 5.1 leads to proper phase space reconstruction. This is guaranteed if m is larger than twice the dimension of the true phase space [1, 6]. However, the true embedding dimension of the attractor is usually unknown in practical applications. The most widely used method for selecting the embedding dimension is based on the concept of *false neighbor*. Let us suppose that we have built a state space reconstruction using an embedding dimension m and a time lag τ (proper time lag selection will be exposed below). Given a specific vector $\mathbf{x}_i$, we shall denote its nearest neighbor in the phase space by $\mathbf{x}_i^{nn}$. If $\mathbf{x}_i^{nn}$ is a true neighbor of $\mathbf{x}_i$, then it should have arrived to a neighborhood of $\mathbf{x}_i$ through the dynamical rules of the system. On the other hand, if $\mathbf{x}_i^{nn}$ is a false neighbor, then it may have arrived to a neighborhood of $\mathbf{x}_i$ because during the phase space reconstruction process, we have projected the d-dimensional attractor into an m-dimensional space, $d > m$. If we have false neighbors in a given reconstruction, by increasing the dimension of the reconstructed phase space, we should decrease the number of false neighbors. When we reach the real dimension, only the real neighbors should remain in the neighborhood of the point, while false neighbors should disappear. Henceforth, increasing the size of the reconstructed space should not decrease the number of neighbors. Thus, the key idea behind the most widely used algorithms for selecting a proper phase space dimension is searching for the minimum dimension in which most of the false neighbors have been eliminated. RHRV implements one of the best-known algorithm based on false neighbo rs [7] (see Sect. 5.4.2).

Selecting a proper time lag parameter τ is also important when embedding the vectors in an m-dimensional phase space. If τ is small compared to the time scale that the system needs to manifest noticeable changes, successive elements of the phase space will be highly correlated. Thus, they will tend to cluster around the diagonal of $\mathbb{R}^m$. On the other hand, if τ is too large, the phase space elements will be independent; thus, the resulting embedding will tend to spread in $\mathbb{R}^m$. The first zero in the autocorrelation function of the signal is often selected as a compromise between these extreme situations. The first minimum and the $1/e$ value of the autocorrelation function are also used as selection criteria for τ.

This procedure is sometimes criticized because the selection of τ relies on a linear statistic (the autocorrelation) and thus, it does not take into account the nonlinear correlations of the process. Fraser et al. suggested the use of the average mutual information (AMI) between the time series and a delayed version of itself for selecting the time lag parameter [8]. The AMI function, $AMI[\tau]$, quantifies the amount of knowledge gained about the value of $x_{i+\tau}$ when observing x_i by studying the probability distribution of $(x_i, x_{i+\tau})$. The first minimum and the first $1/e$ decay of $AMI[\tau]$ are commonly used for selecting the time lag parameter τ. The *CalculateTimeLag* function of RHRV permits the user to compute a proper time lag using either the autocorrelation function or the average mutual information (see Sect. 5.4.2).

In the next sections, a brief theoretical discussion of the nonlinear deterministic techniques implemented in RHRV is presented. For further details, the interested reader is referred to [1, 6].

5.2.3 Correlation Dimension

Some systems with highly complicated dynamics in time also show its complexity in phase space, generating attractors with complicated geometry: the so-called *strange attractors*. These attractors are often fractal sets and they are generated by *chaotic systems*, i.e., systems that despite being completely deterministic are unpredictable in the long term. This has led researchers to try to characterize the dynamical system by measuring the fractal dimension of its attractor. The fractal dimension is a generalization of the Euclidean dimension (they coincide in non-fractal sets) that measures the auto-similarity of the fractal.

There are several techniques that approximate the concept of fractal dimension: the box-counting dimension, the information dimension, the correlation dimension, the Rényi dimension, and the Higuchi dimension. It should be noted that these fractal dimensions are not equivalent to each other. The correlation dimension is the most common measure in the context of dynamical systems [9].

The estimation of the correlation dimension requires the computation of the so-called correlation sum $C(r)$. The correlation sum is defined over the N points from the phase space as follows:

$$C(r) = \frac{2}{N(N-1)} \sum_{i=1}^{N} \sum_{j=i+1}^{N} \Theta(r - \|\mathbf{x}_i - \mathbf{x}_j\|_\infty), \qquad (5.2)$$

where $\Theta(\cdot)$ is the Heaviside step function (which is 0 for negative arguments and 1 for positive arguments) and r is a radius in the phase space. The summation in

Eq. 5.2 just counts the pairs of points $(\mathbf{x}_i, \mathbf{x}_j)$ whose distance in the phase space is smaller than r. Usually, the L_∞ norm is selected in practical implementations; this norm computes the distance between two points as the maximum of the absolute value of the differences between the coordinates of the points. In the remainder of the chapter, distances and neighborhoods should be understood under the L_∞ metric. Note that we could have denoted the correlation sum as $C(m, r)$ to explicitly indicate its dependence on the embedding dimension m used for constructing the vectors $\mathbf{x}_i$. However, to maintain clarity and simplicity in notation, we will omit this dependency when it is not essential.

The correlation sum given by Eq. 5.2 is an estimator of the mean probability of two states (at different times) being close. This estimator is biased when the pairs in the sum are not statistically independent, usually because there is some temporal correlation between Takens' vectors. This issue is usually solved by using the so-called Theiler window: two Takens' vectors must be separated by, at least, the time steps specified by this window in order to be considered neighbors. By using a Theiler window, we exclude temporally correlated vectors from our estimations.

Fractal attractors are expected to fulfill

$$C(r) \propto r^D,$$

being D the correlation dimension. Thus, the correlation dimension can be estimated using a log-log regression of $C(r)$ on r.

The correlation dimension of a fractal attractor is an invariant of the dynamical system. Thus, it should not depend on the dimension of Takens' vectors employed to estimate it (provided that we are using an embedding dimension large enough to reconstruct the phase space). In practical applications, researchers often plot $log(C(r))$ vs. $log(r)$ using several embedding dimensions. To obtain a reliable estimation of the correlation dimension, there must exist a range of r over which $log(C(r))$ is linear in $log(r)$ for all embedding dimensions larger than some m_{min}, being m_{min} the minimum embedding dimension that permits proper phase space reconstruction.

5.2.4 Generalized Correlation Dimension and Information Dimension

As we have previously stated, the correlation sum $C(r)$ can be interpreted as the mean probability of finding a neighbor in a ball of radius r surrounding a point in the phase space: $C(r) = \langle p(r) \rangle$. The average operation is necessary since the number of points in a neighborhood of a fractal attractor is not homogeneous. Furthermore, the average time that the system spends in a region of the phase space is not homogeneous either. The Renyi dimension (or generalized dimension) takes into account the frequency with which the different regions are visited. The strength of the weighting is controlled by a numeric value $q > 0$ (the larger the q, the stronger the importance of the most visited regions). After fixing a n_{min} Theiler window, the generalized correlation dimension is defined:

$$C_q(r) = \frac{1}{(N - 2n_{min})(N - 2n_{min} - 1)^{(q-1)}}$$

$$\times \sum_{i=n_{min}}^{N-n_{min}} \left[\sum_{|j-i|<n_{min}}^{N} \Theta(r - \|\mathbf{x}_i - \mathbf{x}_j\|_\infty) \right]^{q-1},$$

which is an estimation of $\langle p(r)^{(q-1)} \rangle$. For small radius, it is assumed that $\langle p(r)^{(q-1)} \rangle \propto r^{(q-1)D_q}$, which defines the generalized dimension of order q. An estimation of the generalized dimension of order q is then obtained by fitting a log-log regression of $C_q(r)$ on $r^{(q-1)}$. Note that the correlation sum presented in the previous section is just a particular case of the generalized dimension when $q = 2$. The case $q = 1$ leads to the information dimension. The information dimension D_1 can be defined as

$$D_1 = \lim_{r \to 0} \frac{\langle \log p(r) \rangle}{\log(r)}. \tag{5.3}$$

Since $\langle \log p(r) \rangle$ is the average Shannon information needed to specify a point in phase with precision r, the information dimension specifies how the average Shannon information scales with the radius r.

In order to estimate D_1 in practical applications, a different approach than the one suggested by Eq. 5.3 is used. Instead of exploring the scaling behavior of $\langle \log p(r) \rangle$ when r changes, the scaling behavior of the average radius $\langle r(p) \rangle$ that contains a given portion of the total points in the phase space (a "fixed-mass" p) is explored:

$$D_1 = \lim_{p \to 0} \frac{\log p}{\langle \log r(p) \rangle},$$

an estimate of D_1 is usually obtained by performing a linear regression of $\log p$ on $\langle \log r \rangle$.

The considerations discussed in Sect. 5.2.3 for obtaining a reliable correlation dimension estimate are also valid for the generalized dimension and the information dimension. That is, we should compute $C_q(r)$ (or $\langle \log r(p) \rangle$ for information dimension estimates) for several m starting at a reasonable embedding dimension. Only if a clear scaling regions are found for all embedding dimensions larger than some m_{min} can we be confident about the estimations.

5.2.5 Kolmogorov-Sinai Entropy

In 1948, C. Shannon introduced the concept of information entropy to measure the average amount of information present in a message. Intuitively, high entropy means that each character provides new information, whereas low entropy implies that new characters repeat something that is already known. The *Kolmogorov-Sinai entropy* (or KS entropy) is a generalization of the Shannon entropy that arises when considering an observed time series as a transmitted message. The KS entropy measures the unpredictability of a time series. The larger the entropy is, the higher the unpredictability of the system. Note that this is consistent with the intuitive interpretation of the Shannon entropy: if a time series is predictable from the observation of past samples, new values (new characters) will not provide new information.

KS entropy is not suitable for practical applications since it requires large amounts of data and it is sensitive to the presence of noise. The *approximate entropy* [10] and the (closely related) *sample entropy* [11] methods were developed to overcome the limitations of the KS entropy when working with real data. Although RHRV implements both algorithms, the approximate entropy method is now deprecated, and thus, we will focus on the sample entropy. The sample entropy of order order q can be computed by using correlation sums of order q. We first define the function:

$$h_q(m, r) = \log \left[\frac{C_q(m, r)}{C_q(m+1, r)} \right], \tag{5.4}$$

where m is the embedding dimension and q the order of the correlation sum. For sufficiently large m, $h_q(m, r)$ should converge toward a constant H_q, which is defined as the sample entropy of order q. If the sample entropy is properly estimated, the $h_q(m, r)$ function should be approximately constant for all embedding dimensions larger than some m_{min} in some range of r.

5.2.6 Maximal Lyapunov Exponent

The best well-known characteristic of a chaotic system is its sensitivity to initial conditions. This means that any two trajectories starting out very close to each other will separate exponentially fast in time. This sensitivity to initial conditions is usually quantified as

$$\delta(t) \approx \delta(0) \cdot \exp(\lambda t), \tag{5.5}$$

where $\delta(0)$ is the initial distance between two close points in phase space (thus, we assume that $\delta(0)$ is small), $\delta(t)$ is their distance at time t, and λ is the average divergence rate of close trajectories in the system. λ is called the *maximal Lyapunov exponent*.

From Eq. 5.5,

$$\lambda t \approx \log\left[\delta(t)/\delta(0)\right]. \tag{5.6}$$

Thus, to estimate the Lyapunov exponent of a system, a linear regression of $S(t) = \log\left[\delta(t)/\delta(0)\right]$ on t is usually performed. In practical applications, one should check for the existence of a linear region when representing $S(t)$ against t. If for some temporal range this plot shows a linear behavior, its slope is an estimate of the maximal Lyapunov exponent per unit of time. If such a region does not exist, the estimation should be discarded.

Under some general restrictions, the Lyapunov exponent becomes a well-defined invariant quantity of the dynamical system [1]. Thus, just like for the generalized dimension estimates, the maximal Lyapunov exponent should be computed for several embedding dimensions in order to check that it does not depend on the embedding dimension and that the quantity is, indeed, invariant.

5.2.7 Recurrence Quantification Analysis (RQA)

The recurrence quantification analysis (RQA) quantifies the number and duration of the recurrences in the phase space. A recurrence is the return of the trajectory in phase space to a neighborhood of a region that was visited before. RQA requires the computation of the *recurrence matrix*:

$$M_{i,j}(r) = \Theta(r - \|\mathbf{x}_i - \mathbf{x}_j\|_\infty),$$

where $\mathbf{x}_i$ and $\mathbf{x}_j$ are the phase space vectors, r is a small radius that defines the size of the neighborhood, and $\Theta(\cdot)$ is the Heaviside function. Thus, if a point (i, j) is marked as recurrent, this means that the state of the system at time i has some similarity with the state of the system at time j. The degree of similarity is measured through the radius r.

The graphical representation of the recurrence matrix is usually referred to as recurrence plot (RP). The appearance of a RP provides evidence about the dynamical properties of the system. For example, a homogeneous RP indicates stationarity. Furthermore, if the RP is homogeneous in a small block of the plot, a stationary behavior can be assumed for the corresponding time period.

Vertical and horizontal lines indicate a time segment in which a state does not change or changes very slowly. Diagonal lines occur when a segment of a trajectory visits the same region of the phase space at different times. Therefore, parallel and perpendicular lines to the main diagonal are a strong sign of periodicity or determinism. For periodic or almost-periodic signals, the distance between parallel lines corresponds to the period.

5.2 Chaotic Nonlinear Statistics

White bands in the RP are usually due to abrupt changes in the dynamics that lead to rare events. Thus, they indicate non-stationarity.

To quantify the presence of these visual patterns in RPs, Zbilut et al. proposed the use of some simple statistics that can be extracted from the recurrence matrix [12]. The characterization of the recurrence matrix through these statistics is what receives the name of recurrence quantification analysis (RQA).

The simplest measure that can be extracted from the recurrence matrix is the recurrence rate REC:

$$REC = \frac{1}{N^2} \sum_{i,j=1}^{N} M_{i,j}(r),$$

where N is the number of points on the phase space trajectory. The recurrence rate is a measure of the density of the recurrent points in the RP. RQA is usually applied to sparse recurrence matrices, and therefore the recurrence rate must be kept low (e.g., from 0.1 to 2.0%).

The recurrence rate can also be computed separately for each diagonal parallel to the main diagonal of the recurrence matrix at a given distance t. The resulting statistic is referred to as the t-recurrence rate:

$$REC_t = \frac{1}{N-t} \sum_{i=1}^{N-t} M_{i,i+t}(r).$$

The t-recurrence rate may be interpreted as the probability that a state returns to its neighborhood after t steps.

The trend statistic is a useful RQA measure based on the t-recurrence rate. The trend is a linear regression coefficient obtained when considering a linear relation between REC_t and t:

$$TREND = \frac{\sum_{t=1}^{\tilde{N}}(t - \tilde{N}/2)(REC_t - \overline{REC_t})}{\sum_{t=1}^{\tilde{N}}(t - \tilde{N}/2)^2},$$

where $\overline{REC_t}$ is the mean REC_T value and $\tilde{N}$ defines the maximal number of diagonals parallel to the main diagonal which will be used for computing the trend. The trend provides information about the stationarity of the process. In a stationary system, the recurrent points are homogeneously distributed across the RP. Hence, the trend will be near zero.

Weakly correlated processes with stochastic dynamics generate RPs with small diagonal lines and isolated recurrent points, whereas deterministic processes usually generate longer diagonal lines. Thus, the ratio of recurrence points that form diagonals to all the recurrent points is a good measure of determinism. Thus, the determinism parameter is defined as

$$DET = \frac{\sum_{l=l_{min}}^{N} l \cdot P(l)}{\sum_{i,j=1}^{N} M_{i,j}(r)},$$

where $P(l)$ is the number of diagonal segments of length l. The diagonal segments must have a minimum length l_{min}; otherwise, they are excluded.

Another statistic related to diagonal lines is the ratio, defined as the ratio between *DET* and *REC*. The ratio has proved useful for detecting transitions between physiological states [13]. During transitions, the number of recurrent points usually decreases, whereas the proportion of points in diagonal lines is less affected. Hence, the ratio increases during dynamical transitions.

A diagonal line with length l implies that a segment of a trajectory is close to another segment (occurring at a different time) during l time steps. Thus, the length of the diagonal lines depends on the rate of divergence of close trajectories in phase space. The average diagonal line length

$$L = \frac{\sum_{l=l_{min}}^{N} l \cdot P(l)}{\sum_{l=l_{min}}^{N} P(l)}$$

may be interpreted as the mean time that two segments of the trajectory are closed to each other.

Other RQA measures related with diagonal lines are the length of the longest diagonal line L_{max} and its inverse, the divergence $DIV = \frac{1}{L_{max}}$. Since the lengths of the diagonal lines are related to the divergence time of close trajectories, L_{max} and *DIV* are related to the maximal Lyapunov exponent. *DIV* scales proportionally with the maximal Lyapunov exponent, whereas L_{max} scales inversely.

The Shannon entropy of the probability $p(l)$ to find a diagonal line of length l,

$$ENTR = -\sum_{l=l_{min}}^{N} p(l) \log p(l),$$

is a measure of the signal's complexity. A large entropy is typical of chaotic systems, whereas a low entropy indicates either a periodic signal (since all its diagonal lines are long) or an uncorrelated noise (all its diagonal lines are small).

It is possible to define analogous statistics to quantify the presence of vertical lines. The ratio between the recurrent points forming vertical lines and the entire set of recurrent points is called laminarity:

$$LAM = \frac{\sum_{v=v_{min}}^{N} v \cdot P(v)}{\sum_{i,j=1}^{N} M_{i,j}(r)}.$$

5.2 Chaotic Nonlinear Statistics

Table 5.1 Most important RQA statistics

Statistic	RHRV name	Interpretation
Recurrence	REC	Percentage of recurrent points in a recurrence plot
Determinism	DET	Percentage of recurrent points that form diagonal lines. The presence of long diagonal lines is a sign of determinism
Ratio	RATIO	Ratio between DET and REC. That is, the ratio between the number of recurrent points in diagonal lines and the total number of points of the recurrence matrix
Laminarity	LAM	Percentage of recurrent points that form vertical lines. A large value of laminarity indicates the existence of laminar states, in which the system is trapped for some time
Longest diagonal line	Lmax	Length of the longest diagonal line. This quantity inversely scales with the maximal Lyapunov exponent
Divergence	DIV	Inverse of Lmax. This quantity scales with the maximal Lyapunov exponent and thus, it measures the rate of divergence of the system
Averaged diagonal line length	Lmean and LmeanWithout-Main	Mean length of the diagonal lines. This quantity is related with the mean time that two close states in the phase space need to diverge *Lmean* takes into account the main diagonal, whereas *LmeanWithoutMain* does not
Trapping time	Vmean	Average length of the vertical lines. This quantity is related with the mean time that the system spends trapped in some state
Longest vertical line	Vmax	Longest vertical line, which is the maximum time that the system spends trapped in a state
Entropy	ENTR	Shannon entropy of the diagonal line length distribution
t-recurrence rate	recurrenceRate	Percentage of recurrent points in a diagonal line separated t steps from the main diagonal line
Trend	TREND	Trend of the number of recurrent points depending on the distance to the main diagonal

The average length of the vertical lines is referred to as the trapping time:

$$TT = \frac{\sum_{v=v_{min}}^{N} v \cdot P(v)}{\sum_{v=v_{min}}^{N} P(v)},$$

and it is related with the mean time that the system spends trapped in some state.

Finally, the length of the longest vertical line, V_{max}, may be interpreted as the maximum time that the system spends trapped in a state.

Table 5.1 summarizes the interpretation of all the RQA statistics previously discussed and the name that they receive in RHRV (see Sect. 5.4.7). For a more detailed revision of the RQA, the interested reader may consult [12, 14].

5.2.8 Poincaré Plot

The Poincaré plot is a graphical representation of the dependence between successive RR intervals obtained by plotting the $RR_{j+\tau}$ as a function of RR_j, being τ a positive integer. This dependence is often quantified by fitting an ellipse to the plot, which results in two parameters characterizing the ellipse: SD_1 and SD_2. SD_1 is usually calculated as the standard deviation of the points perpendicular to the line of identity, and SD_2 is calculated as the standard deviation along the line of identity. Although Poincaré plots have its origin in nonlinear theory, Brennan et al. demonstrated that SD_1 and SD_2 are incapable of identifying the nonlinear beat-to-beat structure [15]. Indeed, for $\tau = 1$, it can be demonstrated that SD_1 and SD_2 can be related to linear time-domain measures [15]:

$$SD_1^2 = \frac{1}{2} SDSD^2, \tag{5.7}$$

$$SD_2^2 = 2 \cdot SDNN^2 - \frac{1}{2} SDSD^2 \tag{5.8}$$

Thus, SD_1 characterizes short-term variability, whereas SD_2 characterizes long-term variability.

5.3 An Overview of Fractal Dynamics

Fractal time-series model some fundamental features present in RR intervals: self-affinity, power law scaling, and scale invariance. When small segments of self-affine objects are enlarged, each with different scaling factors in each direction, they exhibit statistical similarity to larger counterparts, as illustrated in Fig. 5.1. As a consequence of this scale invariance, when measuring a quantity Q that depends on the scale s, self-affine objects manifest a remarkable property: Q does not converge to a single value, but instead it exhibits a power law scaling relationship:

$$Q \propto s^H, \tag{5.9}$$

and thus, Q depends on the scale of measurement (this property is discussed in the first Mandelbrot's paper about fractals [16]). As a consequence of Eq. 5.9, scale invariance emerges: The ratio of two estimates at different scales only depends on the ratio of the scales:

$$\frac{Q(s_1)}{Q(s_2)} = \left(\frac{s_1}{s_2}\right)^H.$$

5.3 An Overview of Fractal Dynamics

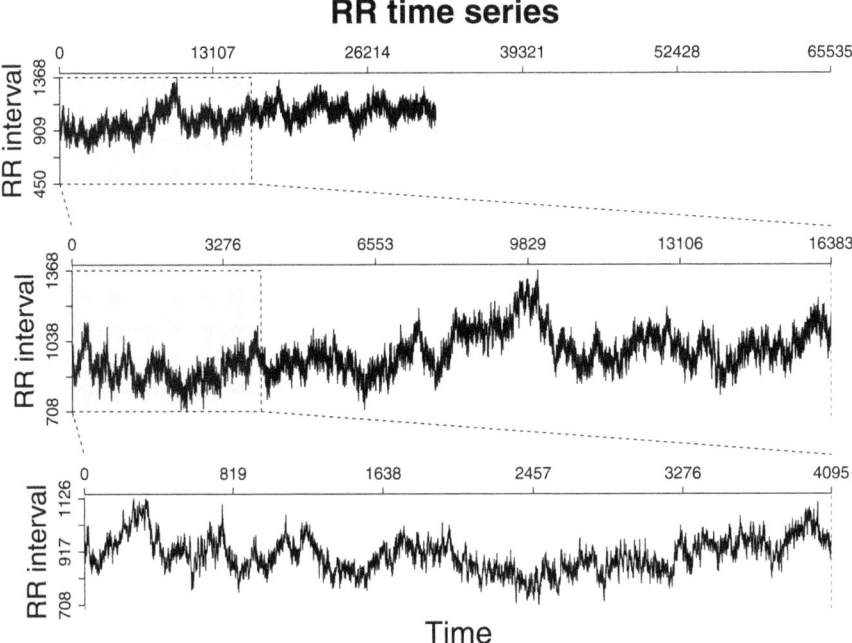

Fig. 5.1 Self-affinity of a simulated RR series following a fractional Brownian motion

Eke et al. [17] proposed a dichotomous model in which physiological signals are seen as realizations of one of two temporal processes: *fractional Brownian motion* (fBm) and *fractional Gaussian noise* (fGn) [17]. The self-affinity and the appearance of both fBm and fGn are controlled by a single parameter, the *Hurst exponent* $0 < H < 1$. fBm signals are nonstationary. A fBm signal with an H close to 1 describes a smooth signal, whereas a fBm signal with H close to 0 describes a rough signal. The increments of a nonstationary fBm yield a stationary fGn. Thus, we can transform a fBm signal in a fGn signal by differencing, and we can transform a fGn signal in a fBm signal by cumulative summation. The H exponent of a fGn can be easily interpreted:

- If $H < 1/2$, the process is anti-correlated (negatively correlated).
- If $H = 1/2$, the process is uncorrelated (white noise). The cumulative summation of white noise yields Brownian motion (fBm with $H = 1/2$).
- If $H > 1/2$, the process is positively correlated.

The next sections provide a brief overview of the most important tools for estimating the Hurst exponent included in RHRV. Note that the Hurst exponent is between 0 and 1 regardless of the model. Thus, to avoid possible mistakes, when reporting values of H, the class of the signal should always be specified (as H_{fBm} or H_{fGn}) [17].

Note that this section focuses on monofractal models. However, there exist evidence supporting the application of *multifractal* models to HRV [18, 19].

5.3.1 Detrended Fluctuation Analysis

Due to its self-affine nature, the standard deviation of a fBm signal in a window of size n, σ_n, fulfills the power law:

$$\mathbb{E}[\sigma_n] \propto n^H. \quad (5.10)$$

Based on Eq. 5.10, Mandelbrot introduced a Hurst estimation technique, named *scaled windowed variance (SWV) analysis* [20]. Peng et al. further developed SWV to improve the analysis of nonstationary data [21]. This method is now referred to as *detrended fluctuation analysis (DFA)*. The DFA algorithm proceeds as follows:

1. The signal to be analyzed x is summed (the mean is subtracted to avoid obtaining a monotonic signal):

$$y_i = \sum_{k=1}^{i}(x_k - \mu).$$

   The resulting time series y is usually referred to as the *profile*.
2. The profile is divided into nonoverlapping segments of length n. Then the local trend $y_{i,n}$ in each of the segments is estimated using least square regression.
3. The average fluctuation of the data relative to the local trend is computed:

$$F_n = \sqrt{\frac{1}{N}\sum_{j=1}^{N}(y_j - y_{j,n})^2},$$

   where N is the number of nonoverlapping segments.
4. Repeat steps 1–3 for different windows sizes to obtain the so-called fluctuation function F_n.
5. If the data is self-affine, $F_n \propto n^\alpha$, and the scaling exponent α is estimated using regression.

The estimated scaling exponent, $\hat{\alpha}$, can be used for discriminating the signal nature. If $\hat{\alpha} < 1$, then the signal is fGn and $\hat{\alpha} = \hat{H}_{fGn}$, whereas that if $\hat{\alpha} > 1$, the signal is fBm with $\hat{\alpha} = \hat{H}_{fBm} + 1$. If $\hat{\alpha} \approx 1$, there is uncertainty about the nature of the signal.

5.3.2 Power Spectral Density Analysis

The power spectrum of a fractal signal fulfills the power law relationship:

$$S(\omega) = C\omega^{-\beta}, \tag{5.11}$$

where β is the *spectral index*. Signals with this type of power spectrum are referred to as *1/f noises*. 1/f noises with $-1 < \beta < 1$ and $1 < \beta < 3$ can be approximated by fGn and fBm signals, respectively [17].

From Eq. 5.11, it is clear that we may estimate β using regression. The high-frequency part of the spectrum is usually discarded before fitting the regression line, since in practical applications, the power law behavior holds only for frequencies close to zero [22]. The resulting estimation, $\hat{\beta}$, can be used for discriminating the signal class. If $\hat{\beta} < 1$, then the signal is a fGn, whereas if $\hat{\beta} > 1$, the signal is a fBm. If $\hat{\beta} \approx 1$, there is uncertainty about the nature of the signal. Once the signal has been classified, we may obtain the Hurst exponent using

$$\hat{\beta} = 2\hat{H}_{fBm} + 1 \quad \text{or} \quad \hat{\beta} = 2\hat{H}_{fGn} - 1.$$

5.4 Chaotic Nonlinear Analysis with RHRV

To illustrate the nonlinear and fractal analysis techniques introduced in Sects. 5.2 and 5.3, we shall study two segments of a recording from a patient suffering from obstructive sleep apnea (OSA). OSA is a sleep disorder characterized by breathing cessation during the nocturnal rest. As a consequence of breathing interruption, the patient's sleep can be severely disrupted. Apnea patients do not usually remember the breathing cessations, but they feel fatigued and sleepy during daytime. If apnea is not properly treated, it can result in several cardiovascular diseases, such as hypertension, Type II diabetes, cardiovascular disease, and stroke [23]. The recording we shall work with is the "a05" file from the PhysioNet Apnea-ECG database. This database is described in [24] and the recording can be downloaded from [25]. Since this recording contains both apnea and normal-resting episodes, we will focus on two small segments in which the time series looks stationary: a segment in which the patient is suffering from apnea and another one in which the patient is breathing normally. To extract a subset of the RR time series from a *HRVData* structure, the *ExtractTimeSegment* function can be used. There also exists an alias for this function named *Window*. In addition to the *HRVData* structure, the *ExtractTimeSegment* function takes as input parameters the following:

- *starttime* and *endtime*: which indicate the start and end times of the period of interest. *ExtractTimeSegment* extracts the RR intervals observed between the times *starttime* and *endtime*. If the *HRVData* contains the interpolated HR series, *ExtractTimeSegment* also returns an interpolated HR series within the interval of interest. However, all analyses of the original *HRVData* will not be part of the extracted data.

Listing 5.4.1 makes use of the *ExtractTimeSegment* function to extract the segments of interest from the "a05" recording. Listing 5.4.1 also plots a small detail of the resulting series (see Fig. 5.2). During the apnea episode (top of Fig. 5.2), the RR series exhibits a complex oscillatory pattern in which several frequencies are present. The deterministic nonlinear dynamic is apparent from the deformed wave profile with pointed peaks. These cyclic patterns in HR often appear during prolonged OSA episodes and the resumption of breathing. The blood oxygen desaturation level plays a fundamental role in the mechanism that produces these oscillations [26].

R listing 5.4.1

```
hrv.data <- CreateHRVData()
hrv.data <- LoadBeatWFDB(hrv.data, RecordName = "a05",
                        RecordPath = "./sampleData/",
                        annotator = "qrs")
hrv.data <- BuildNIHR(hrv.data)
hrv.data <- FilterNIHR(hrv.data)
hrv.data <- SetVerbose(hrv.data, T)
hrv.apn <- ExtractTimeSegment(hrv.data, 10500, 12300)
hrv.norm <- ExtractTimeSegment(hrv.data, 22000, 24500)
par(mfrow = c(2, 1))
PlotNIHR(hrv.apn, xlim = c(10500, 11000),
         main = "Apnea episode")

## Plotting non-interpolated instantaneous heart rate
## Number of points: 1976

PlotNIHR(hrv.norm, main = "Normal breathing")

## Plotting non-interpolated instantaneous heart rate
## Number of points: 2640
```

While during long periods of OSA heart rate shows cyclic oscillations, during normal respiration, heart rate dynamics exhibit a broadband, inverse power law spectral distribution [27]. The bottom of Fig. 5.2 also suggests that a stochastic model is more appropriate than a deterministic one. Thus, in Sect. 5.5, we will use fractal analysis techniques to characterize the dynamics of the "a05" recording while the patient is breathing spontaneously.

In the next sections, we will analyze the apnea segment (stored in *hrv.apn*) using nonlinear analysis techniques. However, before starting to work on a nonlinear analysis, we must create a data structure that will store the results of the analysis. Each nonlinear analysis is identified by the position at which it is stored under the *NonLinearAnalysis* list (see Fig. 2.2). To create the nonlinear analysis structure, the *CreateNonLinearAnalysis* function can be used (see Listing 5.4.2).

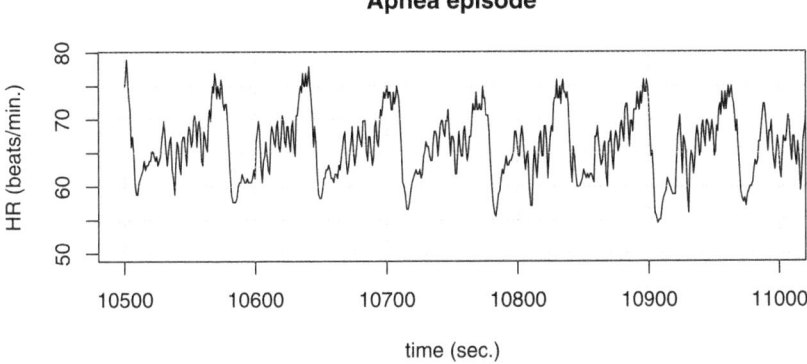

Fig. 5.2 RR time series while the patient is suffering from obstructive apnea (top) and during normal breathing (bottom)

5.4.1 Nonlinearity Tests

First, we should check that the RR intervals during the apnea episode show some degree of nonlinearity (see Sect. 5.2.1). The *SurrogateTest* function of RHRV generates an ensemble of surrogates under the assumption that the data is a stationary linear Gaussian stochastic process. Thus, each of the surrogates has the same mean and autocorrelation as the observed time series. After generating the surrogate data, the *SurrogateTest* function will compute some statistic (specified by the user) for each of the surrogates, in order to test the null hypothesis. The *SurrogateTest* function accepts as input parameters the following:

- *indexNonLinearAnalysis*: numerical value that indicates in which nonlinear analysis structure results will be stored. The default value uses the last analysis structure that has been created.

- *significance*: significance of the test, α (default: 0.05).
- *oneSided*: logical value. If *TRUE*, the routine runs a one-sided test. If *FALSE*, a two-sided test is applied.
- *alternative*: specifies the particular type of one-sided test that should be performed, i.e., if the user wants to test if the statistic from the original data is smaller (*alternative*="*smaller*") or larger (*alternative*="*larger*") than the expected value under the null hypothesis.
- *K*: integer controlling the number of surrogates to be generated. Larger values of K give a more sensitive test than $K = 1$.
- *doPlot*: logical value. If *TRUE*, a graphical representation of the statistics obtained for the surrogates and the original data is shown.
- *useFunction*: an R function computing the discriminating statistic that shall be used for testing. Additional arguments for this function may be specified through the special ... argument.

As an illustrative example, let us consider the higher-order statistic:

$$\langle RR_i \cdot RR_{i+1}^2 - RR_i^2 \cdot RR_{i+1} \rangle.$$

This statistic is often used for detecting nonlinearity, since it measures time asymmetry, and linear stochastic processes are symmetric under time reversal. Listing 5.4.2 implements this statistic and uses it for surrogate data testing (using a two-sided test). From Fig. 5.3, it is clear that we can reject the null hypothesis of a linear stochastic stationary process.

R listing 5.4.2

```
hrv.apn <- CreateNonLinearAnalysis(hrv.apn)

## Creating non linear analysis
## Data has now  1  nonlinear analysis

asymmetryStatistic <- function(x){
  x.len = length(x)
  mean(x[1:(x.len - 1)] * x[2:(x.len)] ^ 2 -
       x[1:(x.len - 1)] ^ 2 * x[2:(x.len)])
}
hrv.apn <- SurrogateTest(hrv.apn, oneSided = F,
                        significance = 0.01, K = 5,
                        useFunction = asymmetryStatistic,
                        doPlot = TRUE)

## Computing statistics
## Null Hypothesis: Data comes from a linear stochastic process
## Reject Null hypothesis:
##  Original data's stat is significant larger than surrogates' stats
```

Surrogate data testing is a complex topic that is beyond the scope of this book. However, we would like to stress that a rejection at a given significance level α does not imply that the data is indeed nonlinear. The rejection may be due to the

5.4 Chaotic Nonlinear Analysis with RHRV

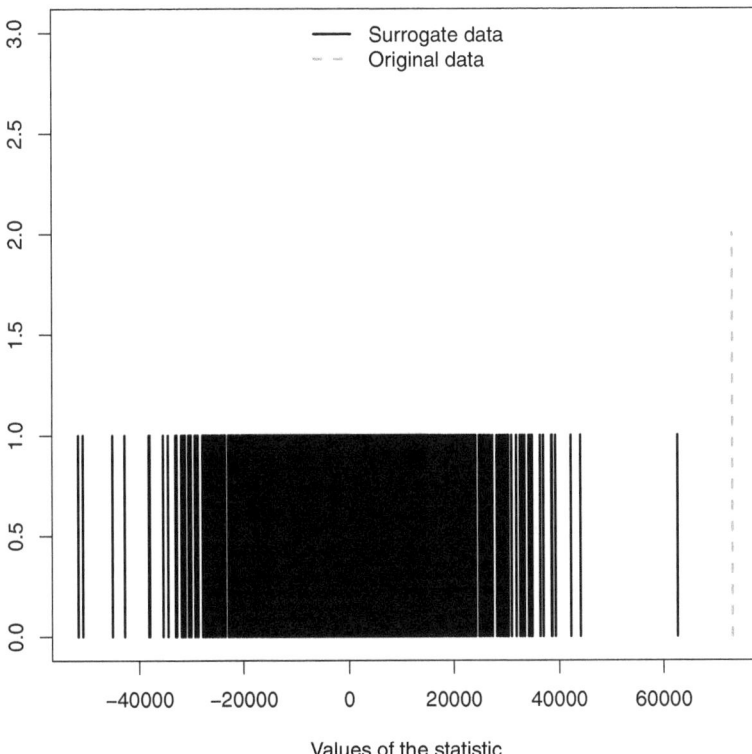

Fig. 5.3 Time asymmetry of the RR data and 199 surrogates. We can reject the null hypothesis of a linear stochastic stationary process

fact that the data was nonstationary or that it was generated from a non-Gaussian process. The rejection could even have happened by chance (with a probability of α). Finally, if we run more than just one surrogate test, we must pay attention to the significance level. For example, five independent tests at the 95% level each yield a significance of 23% (since the probability of false rejection is equal to the probability of rejecting one or more of these independent tests: $1 - 0.95^5 \approx 0.23$). A conservative approximation that always leaves room for further tests is to use $\alpha_i = \alpha/2^i$ in the i-th test when α is the desired significance level. For further details about these topics, the interested reader is referred to [2, 3].

5.4.2 Phase Space Reconstruction

Once we have checked that the data is compatible with a nonlinear process, we must build the phase space from the RR series (see Sect. 5.2.2). The first step that is usually performed in order to achieve phase space reconstruction through the embedding theorem (see Eq. 5.1) is selecting the time lag parameter τ.

The *CalculateTimeLag* function permits the user to compute a proper time lag using either the autocorrelation function (ACF) or the average mutual information (AMI). This function takes as input parameters the following:

- *technique*: the technique to be used to estimate the time lag. Allowed values are *"acf"* and *"ami"*. Depending on the selected technique, additional parameters can be specified. The ACF computations are based on the *acf* function (stats package), whereas the AMI computations are based on the *mutualInformation* function (nonlinearTseries package).
- *method*: the criterion to select a particular time lag. Available methods are *"first.zero"* (selects the time lag where the ACF/AMI function decays to 0), *"first.e.decay"* (time lag is selected where the ACF/AMI decays to 1/e of its value at lag 0), *"first.minimum"* (time lag is selected at the first minimum of the ACF/AMI function), and *"first.value"* (selects the time lag where the ACF/AMI decays to some value specified by the user). It must be noted that the *"first.zero"* method is not appropriate when using the AMI, since it only takes positive values.
- *value*: numeric value indicating the value that the autocorrelation/AMI function must cross in order to select the time lag. It is only used with the *"first.value"* selection method.
- *lagMax*: maximum lag at which to calculate the ACF/AMI.
- *doPlot*: logical value. If *TRUE*, a plot of the autocorrelation/AMI function is shown. Typical graphical parameters can be specified by the user (*main*, *xlab*, ...).

Listing 5.4.3 illustrates the use of the *CalculateTimeLag* function when employing the ACF. Note that the first call is using default values for all the parameters, so we could have not specified them. The plot resulting from this first call is shown in Fig. 5.4. In this case, the *"first.minimum"* method does not yield a proper time lag parameter: the autocorrelation of -0.2 (at time lag 19, see Fig. 5.4) is, for our purposes, equivalent to an autocorrelation of 0.2, which is reached at time lag 7. Thus, basing our decision only on the ACF, we may select any time lag between 7 and 11. However, it is worth using the AMI before selecting a final time lag parameter.

5.4 Chaotic Nonlinear Analysis with RHRV

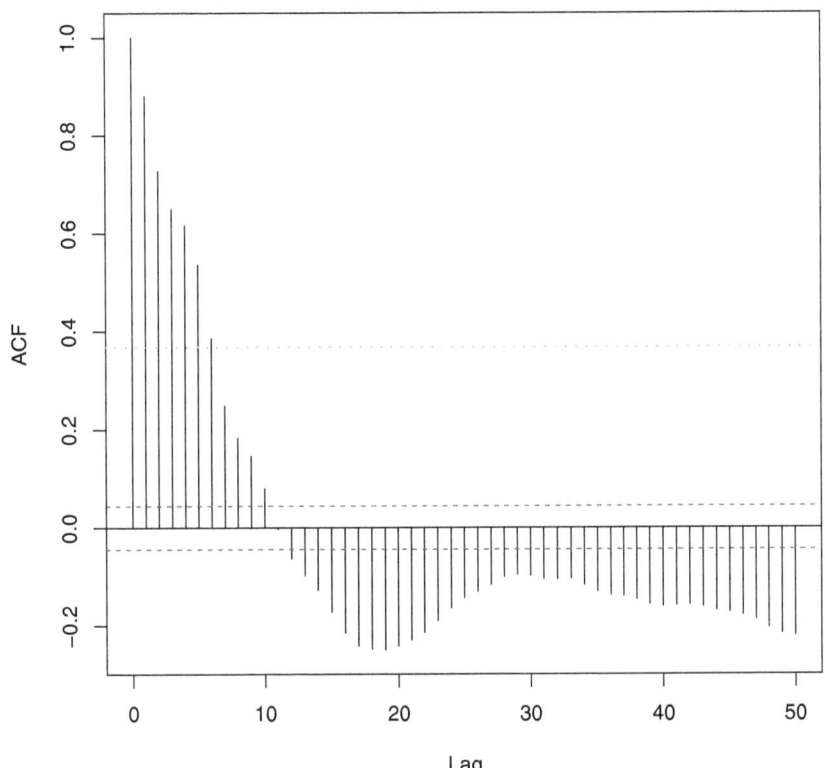

Fig. 5.4 Time lag selection using the ACF. The dotted horizontal line marks the threshold value used for selecting the time lag

R listing 5.4.3

```
# This call is equivalent to CalculateTimeLag(hrv.apn)
time.lag <- CalculateTimeLag(hrv.apn, technique = "acf",
                             method = "first.e.decay",
                             lagMax = 50, doPlot = TRUE)

## Calculating optimum time lag
## Time Lag = 7

time.lag <- CalculateTimeLag(hrv.apn, technique = "acf",
                             method = "first.zero",
                             lagMax = 20, doPlot = FALSE)

## Calculating optimum time lag
## Time Lag = 11

time.lag <- CalculateTimeLag(hrv.apn, technique = "acf",
```

```
                          method = "first.value",
                          value = 0.1, lagMax = 20,
                          doPlot = FALSE)
## Calculating optimum time lag
## Time Lag = 10
time.lag <- CalculateTimeLag(hrv.apn, technique = "acf",
                          method = "first.minimum",
                          lagMax = 20, doPlot = FALSE)
## Calculating optimum time lag
## Time Lag = 19
```

When the AMI is used for selecting the time lag, the following additional parameters may be used:

- *n.partitions*: number of bins used to compute the probability distribution of the time series.
- *units*: the units in which we want to compute the mutual information. Allowed values are *"Nats"*, *"bits"*, and *"Bans"*.

Listing 5.4.4 illustrates the use of the *CalculateTimeLag* with the AMI technique. It must be noted how we avoided the use of the *"first.zero"* selection method, since $AMI[\tau] \geq 0$. Note that the first two calls do not specify the *method* parameter and thus, the *"first.e.decay"* is used. The resulting plot from these two function calls is shown in Fig. 5.5. This figure illustrates how using different numbers of partitions for estimating the probability distribution of $(RR_i, RR_{i+\tau})$ results in small changes in the scale of the AMI. The issue of selecting the number of partitions is similar to the issue of selecting the number of bins in a histogram. Thus, as a rule of thumb, we may set the number of partitions to the number of bins that yields a nice histogram. If *n.partitions* is not specified, the square root of the number of data points is used (a widely used rule for selecting the bins in a histogram). However, since we are more interested in the dependence of $AMI[\tau]$ on τ than in its absolute values, this slight scale variability is not a big issue. Note that, according to the *"first.e.decay"* method, a time lag $\tau = 1$ should be selected. However, taking into account the behavior of $AMI[\tau]$ and the time lags that the use of the ACF suggests (time lags between 7 and 11), it is worth selecting $\tau = 7$ as a compromise. We store the selected time lag in the R variable *selectedTL* for future uses.

The ACF and AMI calculations can also be used to select a proper Theiler window (Sect. 5.2.3). Given that we have a short time series, we should select the minimum window that ensures small temporal correlation between Takens' vectors. A reasonable selection based on Figs. 5.4 and 5.5 is setting the Theiler window to 30 (note that is high enough when compared to the time lag and it also corresponds with two local minima in Figs. 5.4 and 5.5). We store our selection in the R variable *selectedTheiler* for future uses.

5.4 Chaotic Nonlinear Analysis with RHRV

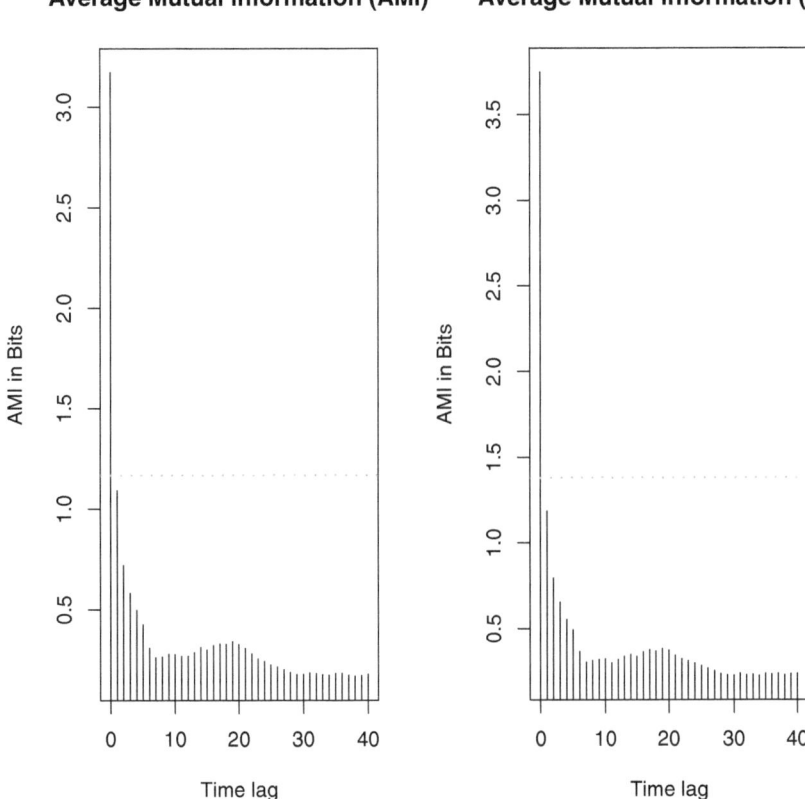

Fig. 5.5 Time lag selection using the AMI. The dotted horizontal lines mark the threshold value used for selecting the time lag. Note the different scales in the y-axis due to the use of different numbers of bins

R listing 5.4.4

```
par(mfrow = c(1, 2))
CalculateTimeLag(hrv.apn, technique = "ami",
                lagMax = 40, doPlot = TRUE,
                units = "Bits")
## Calculating optimum time lag
## Time Lag = 1
   ## [1] 1
CalculateTimeLag(hrv.apn, technique = "ami",
                n.partitions = 18, units = "Bits",
                lagMax = 40, doPlot = TRUE)
## Calculating optimum time lag
## Time Lag = 1
```

```
## [1] 1
CalculateTimeLag(hrv.apn, technique = "ami",
                 method = "first.value", value = 0.25,
                 lagMax = 40, doPlot = FALSE)

## Calculating optimum time lag
## Time Lag = 6

## [1] 6

CalculateTimeLag(hrv.apn, technique = "ami",
                 method = "first.minimum", lagMax = 40,
                 doPlot = FALSE)

## Calculating optimum time lag
## Time Lag = 7

## [1] 7
```

Once we have selected a proper time lag parameter τ, the dimension m of the embedding can be selected using the *CalculateEmbeddingDim* function, which implements Cao's algorithm [7]. Cao's algorithm is one of the most widely used methods for determining the embedding dimension based on the idea of false neighbors (see Sect. 5.2.2).

Cao's algorithm uses two functions to select the optimal embedding dimension. We shall denote this functions by $E_1(d)$ and $E_2(d)$, where d denotes the dimension employed to build Takens' vectors and search for false neighbors. $E_1(d)$ stops changing when d is greater than or equal to the optimal embedding dimension m, staying close to 1. On the other hand, $E_2(d)$ is useful for distinguishing deterministic signals from stochastic signals. For deterministic signals, there exists some d fulfilling $E_2(d) \neq 1$, whereas for stochastic signals, $E_2(d)$ is approximately 1 for all dimensions.

When using the *CalculateEmbeddingDim* function, the user may specify the following input parameters:

- *numberPoints*: number of points from the RR time series used for estimating the embedding dimension. Given that the algorithm requires heavy computations and it may be slow, this parameter can be used to control the computational burden.
- *timeLag*: the time lag used for building Takens' vectors (see Eq. 5.1).
- *maxEmbeddingDim*: maximum dimension at which the E_1 and E_2 functions will be computed. The default value is 15, which is usually enough.
- *threshold*: numerical value that establishes a threshold for considering that $E1(d)$ is close to 1. The default value is 0.95.
- *maxRelativeChange*: maximum relative change in $E_1(d)$ with respect to $E_1(d-1)$ in order to consider that the E_1 function has stabilized and that it has reached the optimum embedding dimension.
- *doPlot*: logical value. If *TRUE* (default), a plot of $E_1(d)$ and $E_2(d)$ is shown.

5.4 Chaotic Nonlinear Analysis with RHRV

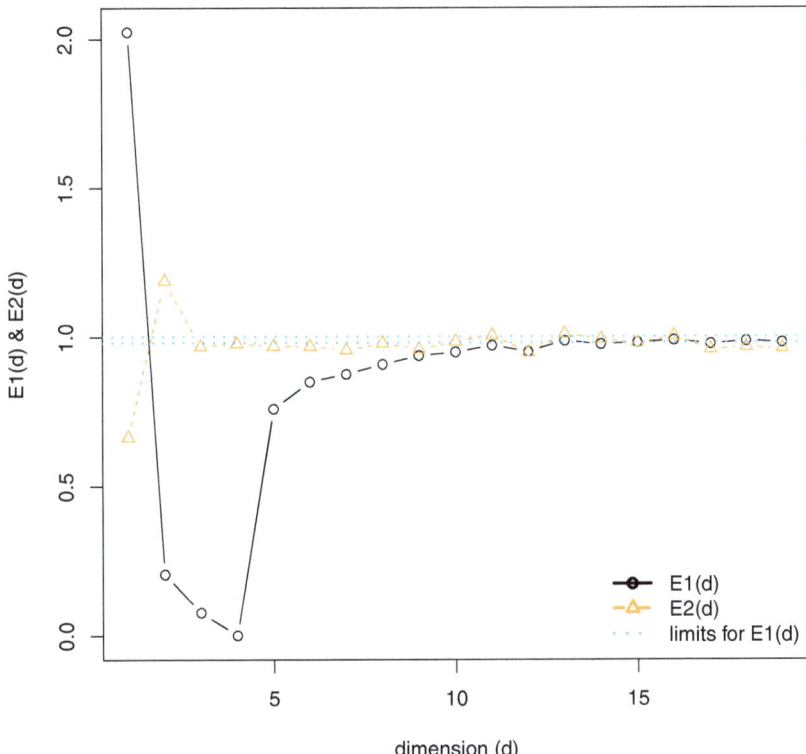

Fig. 5.6 Estimation of the optimum embedding dimension

Listing 5.4.5 makes use of the *CalculateEmbeddingDim* to obtain an estimation of a proper phase space dimension. Note that the *timeLag* parameter has been set to the value estimated in Listings 5.4.3 and 5.4.4. The resulting plot is shown in Fig. 5.6. The first four embedding dimensions have abnormal values for the $E_1(d)$ function. Usually, $E_1(d)$ is an increasing function until it saturates near 1. This strange behavior is due to the finite resolution of the RR series. Indeed, we could have anticipated this effect: the sampling frequency of the ECG from which the RR were derived is 100 Hz, a very low sampling frequency for performing HRV studies [28]. Thus, due to the coarse discretization of the RR series, there exist several identical Takens' vectors for small embedding dimensions. This fact biases the calculation of the $E_1(d)$ since it is not prepared for handling too many identical Takens' vectors. Note that in a deterministic system, the existence of identical phase space vectors implies that the system performs a periodic motion. Since we are interested in complex nonlinear dynamics, we do not consider this kind of simple systems. Aside from the atypical behavior of $E_1(d)$ for small dimensions, $E_1(d)$ behaves as expected and thus, we can trust the estimation.

R listing 5.4.5

```
hrv.apn <- SetVerbose(hrv.apn, T)
selectedED <- CalculateEmbeddingDim(HRVData = hrv.apn,
                                    maxEmbeddingDim = 20,
                                    threshold = 0.98,
                                    timeLag = selectedTL,
                                    numberPoints = 5000)
## Estimating embedding dimension
## Embedding Dimension = 13
```

Once the parameters τ and m have been selected, Eq. 5.1 is used for reconstructing phase space. As a naive check of the resulting phase space reconstruction, we can plot a 3D projection of the attractor that we have obtained as illustrated in Listing 5.4.6. Listing 5.4.6 uses the *BuildTakens* function for obtaining Takens' vectors of the non-interpolated RR intervals. This function takes as input parameters the following:

- *embeddingDim*: the dimension in which we shall embed the RR series
- *timeLag*: the time lag used for building Takens' vectors

The resulting 3D projection of the attractor is shown in Fig. 5.7. The existence of structure in the figure clearly indicates determinism.

R listing 5.4.6

```
library(plot3D)
takens <- BuildTakens(hrv.apn,
                      3, selectedTL)
scatter3D(takens[, 1], takens[, 2], takens[, 3],
          main = "3D projection of the HR attractor",
          phi = 20, theta = 30,
          type = "l", col = 1)
```

5.4.3 Nonlinear Statistics Computation

In practical applications, most of the nonlinear statistics presented in Sect. 5.1 share a common estimation process. This common estimation process is evident when using RHRV, since it employs the same naming convention for computing all nonlinear statistics:

5.4 Chaotic Nonlinear Analysis with RHRV

3D projection of the HR attractor

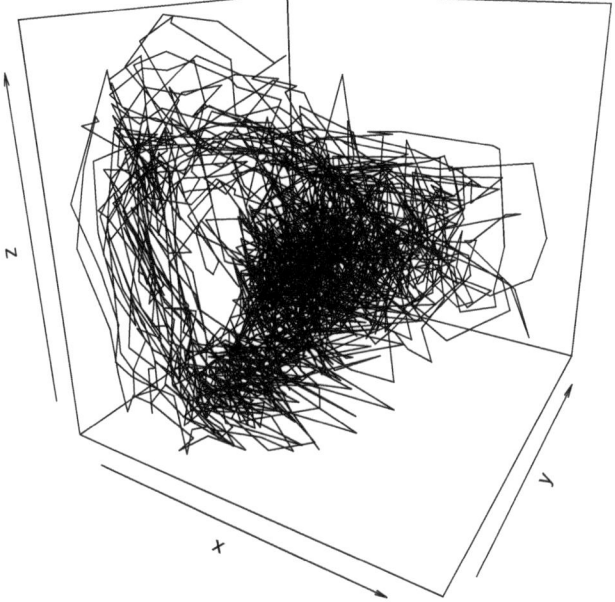

Fig. 5.7 3D projection of the attractor reconstructed from the RR intervals

1. The *CalculateX* function performs some heavy computations characterizing either the scaling behavior of the attractor in the phase space (e.g., correlation dimension) or the dynamical evolution of the system in time (e.g., Lyapunov exponent).
2. The estimation of the nonlinear statistic requires the existence of a small region (in space or time) in which the function computed with *CalculateX* manifests a linear behavior. Thus, it is important to check for the existence of these linear regions through plots. For this purpose, the *doPlot* parameter of the *CalculateX* function or the *PlotX* function can be used.
3. Once the linear region has been localized, the nonlinear statistic is obtained by performing a linear regression using the *EstimateX* function.

In the next sections, we will show how these three steps are performed for estimating the correlation dimension, the sample entropy, and the maximal Lyapunov exponent.

5.4.4 Generalized Correlation Dimension and Information Dimension

The *CalculateCorrDim*, *PlotCorrDim*, and *EstimateCorrDim* are used for estimating any generalized correlation dimension D_q for $q > 1$. In order to compute D_1 (information dimension), the *CalculateInfDim*, *PlotInfDim*, and *EstimateInfDim* functions must be used.

Both *CalculateCorrDim* and *CalculateInfDim* share common parameters. When using them, the user may specify the following:

- *indexNonLinearAnalysis*: numerical value that indicates in which nonlinear analysis structure results will be stored. The default value uses the last analysis structure that has been created.
- The minimum and maximum dimensions in which we will embed the RR series: *minEmbeddingDim* and *maxEmbeddingDim*. The time lag parameter used to construct Takens' vectors is specified by the *timeLag* parameter.
- The size of the Theiler window (see Sect. 5.2.3) can be specified with the *theilerWindow* parameter.
- *doPlot*: if *TRUE*, a plot of the generalized correlation sum or the radius containing a fixed mass of the points (used in the information dimension algorithm) is shown.

Depending on the fractal dimension to be computed, there are also some different parameters that can be specified. These parameters are primarily used for delimiting the range in which the *CalculateCorrDim* and *CalculateInfDim* perform their computations. In the case of the *CalculateCorrDim*, the following parameters can be tuned:

- *CorrOrder*: the order of the correlation sum to compute (denoted with q in the equations).
- The range of radius in which to compute the correlation sum. This range is fully determined by the minimum radius (*minRadius*), the maximum radius (*maxRadius*), and the number of points to estimate (*pointsRadius*). When selecting the range, it must be taken into account that the *CalculateCorrDim* function uses the RR intervals in milliseconds and thus, the *minRadius* and *maxRadius* must be specified in the same unit.

As described in Sect. 5.2.4, the information dimension algorithm uses a different approach than the correlation dimension algorithm, since it looks for the scaling behavior of the average radius that contains a given portion (a "fixed mass" p) of the total points in the phase space. Thus, the "independent" variable is the fixed mass and the user can specify its range. Additionally, to ease the search of the mean radius for each p, the user can define an initial radius for searching for neighbor points in the phase space. If no enough neighbors to reach a given fixed mass are found within this radius, it will be increased by a small factor that the user can also

5.4 Chaotic Nonlinear Analysis with RHRV

tune. Thus, the *CalculateInfDim* function takes as additional input parameters the following:

- The range of the fixed mass, specified by its minimum (*minFixedMass*), its maximum (*maxFixedMass*), and the number of points (*numberFixedMassPoints*).
- *radius*: the initial radius for searching for neighbors in the phase space. Although its choice is not critical, ideally it should be small enough so that the fixed mass contained in this radius is slightly greater than the *minFixedMass*.
- *increasingRadiusFactor*: the factor by which the radius is increased when a given fixed mass is not reached. Note that *increasingRadiusFactor* should be greater than 1.

Listing 5.4.7 computes and plots the correlation sum of the fragment of the "a05" recording where the patient was suffering from apnea. Note that, unlike most RHRV functions, the *CalculateCorrDim* function (and most nonlinear analysis techniques) requires specifying quite a few parameters. Actually, we could have not specified the parameters related to Takens' vectors (time lag and embedding dimensions). In this case, the parameters will be estimated again using the same techniques that we employed in Sect. 5.4.2. However, to avoid unnecessary computations, we specify these parameters. Note that, since we want to verify that the correlation dimension is indeed an invariant of the system (see Sect. 5.2.3), we compute the correlation sum for several embedding dimensions starting from the one selected using Cao's algorithm. The *minRadius* parameter was set to the resolution of the RR intervals: 10 ms (since they were derived from a 100 Hz ECG). On the other hand, Fig. 5.2 shows that the dynamic range of the HR signal is approximately $80 - 55 \approx 25$ beats per minute or, in terms of the RR intervals, 240 ms. To make sure that we cover the whole attractor, the *maxRadius* is set to 300.

R listing 5.4.7

```
hrv.apn <-
  CalculateCorrDim(hrv.apn, timeLag = selectedTL,
                   theilerWindow = selectedTheiler,
                   minEmbeddingDim = selectedED,
                   maxEmbeddingDim = selectedED + 5,
                   minRadius = 10, maxRadius = 300,
                   pointsRadius = 100,
                   doPlot = FALSE)

## Computing the Correlation sum

# magnify axis
PlotCorrDim(hrv.apn,
            cex.lab = 1.4, cex.axis = 1.4,
            cex.main = 1.4, cex.legend = 1.4)
```

As shown in Fig. 5.8, the most prominent feature of the resulting correlation sum is the " staircase" behavior at the lower scales, probably due to the finite resolution of the data. After the identical phase space neighbor issue (Sect. 5.4.2),

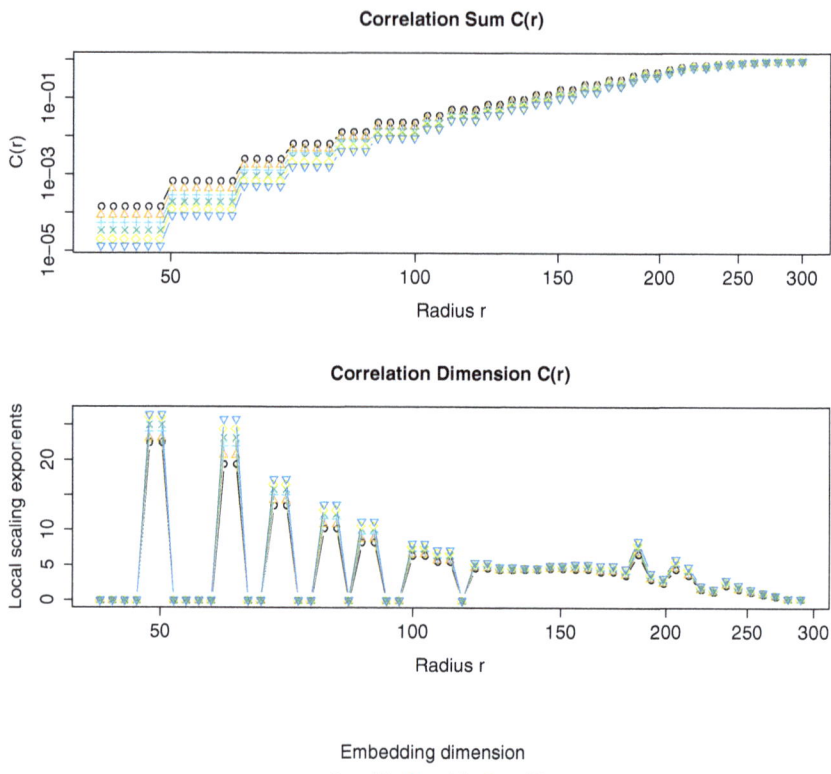

Fig. 5.8 Correlation sum from the "a05" recording. The "staircase" behavior is due to the discretization error

this is the second time that we have encountered a problem caused by the coarsely discretization of the RR intervals. Due to the existence of this problems, when analyzing time series data that has been coarsely discretized, it is recommended to add white noise with an amplitude of the order of the digitizing precision [29]. Then, a nonlinear noise reduction algorithm can be used to restore part of the lost information due to discretization [1].

The *NonLinearNoiseReduction* function implements this procedure. The noise reduction algorithm is based on averaging each Takens' vector in an m-dimensional space with its neighbors (in this case, the time lag parameter is always set to 1). In addition to the *HRVData* structure, *NonLinearNoiseReduction* takes as input the following parameters:

- *embeddingDim*: the dimension in which we shall embed the RR intervals to construct Takens' vectors.

5.4 Chaotic Nonlinear Analysis with RHRV

- *radius*: the radius used to search for neighbors in the phase space and perform averaging. The L_∞ norm is assumed.
- *ECGsamplingFreq*: the sampling frequency of the ECG from which the RR time series was derived. The sampling frequency is used to generate white noise with an amplitude of the order of the digitizing precision. If this parameter is not specified, the ECG resolution is estimated from the RR data.

Listing 5.4.8 illustrates the use of the *NonLinearNoiseReduction* function. Note that although we could have specified the ECG sampling frequency (100 Hz, which means that the resolution is 10 milliseconds), we let the function automatically estimate the ECG resolution. The estimated resolution coincides with the true one. In Listing 5.4.8, the radius is set to *NULL*, and thus, a radius that only depends on the resolution of the RR intervals is automatically selected. Figure 5.9 shows the resulting time series for different values of the *radius* parameter.

R listing 5.4.8

```
hrv.apn <-
  NonLinearNoiseReduction(HRVData = hrv.apn,
                          embeddingDim = selectedED,
                          radius = NULL)

## Denoising RR time series using nonlinear techniques
## Estimated ECG resolution is  10  ms
## Calculating non-interpolated heart rate
## Number of beats: 1976
```

For the remainder of the chapter, we will work with the RR intervals obtained after running Listing 5.4.8. Although the differences between the original RR intervals and those obtained after running the noise reduction algorithm are almost imperceptible (see Fig. 5.9), the "staircase" behavior has disappeared, as shown in Fig. 5.10. We could have used the nonlinear noise reduction algorithm after encountering the atypical behavior of $E_1(d)$ in Sect. 5.4.2, but we postponed its use in order to illustrate the effect of poor discretization in the correlation sum. We suggest the reader to run again Listing 5.4.5 to check that $E_1(d)$ behaves as expected.

Once the discretization issue has been addressed, we may focus on other properties of the correlation sum that can be appreciated in Fig. 5.10. We may distinguish four different regions in the correlation sum. In the *macroscopic regimen* (radius > 150), the scaling behavior is destroyed due to the finite size of the attractor.

On smaller radius, the *scaling region* in which we are interested appears (110 < radius < 150). In a true scaling region, the slope of the correlation sum (the scaling exponent) is constant for all embedding dimensions larger than some m_{min}. If the scaling region is large enough, we may use the scaling exponent as an estimation of the correlation dimension. In Fig. 5.10, the scaling region is quite narrow, and thus, we should not claim with complete certainty to have computed the correlation dimension of the attractor.

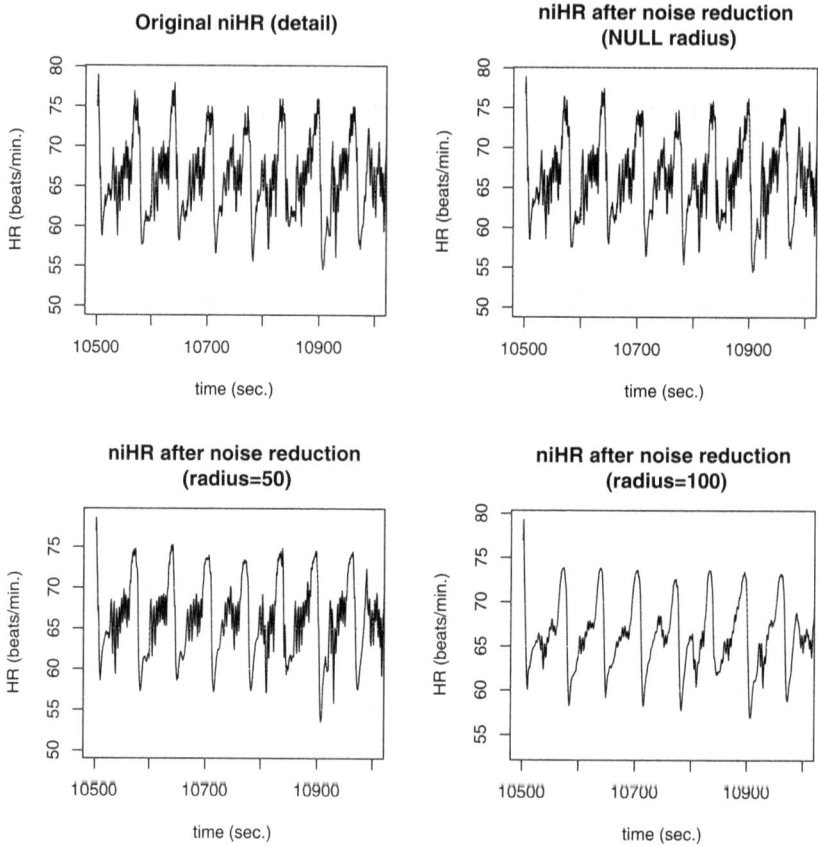

Fig. 5.9 Illustration of the nonlinear noise reduction algorithm and the impact of the radius parameter

On smaller scales, the *noise regime* appears (40 < radius < 100). In this scales, Takens' vectors are not confined to the attractor anymore, but they spread over the whole phase spaced due to noise effects. Thus, the local scaling exponent is close to the embedding dimension m in this range of scales (note the value of the scaling exponent for $r = 50$ in Fig. 5.10). Finally, we may find the issue of lack of neighbors in the smallest scales (radius < 40). In this region, heavy oscillations in the local scaling exponent occur.

Although we cannot be completely confident about the width of the scaling region, we shall estimate the correlation dimension using the *EstimateCorrDim* function for illustrative purposes. This function estimates the scaling exponent of the RR intervals averaging the slopes of the embedding dimensions in which the scaling region is clear. As discussed in Sects. 5.1 and 5.4.3, several nonlinear statistics are estimated using a linear regression in some scaling region. Thus, all *EstimateX*

5.4 Chaotic Nonlinear Analysis with RHRV

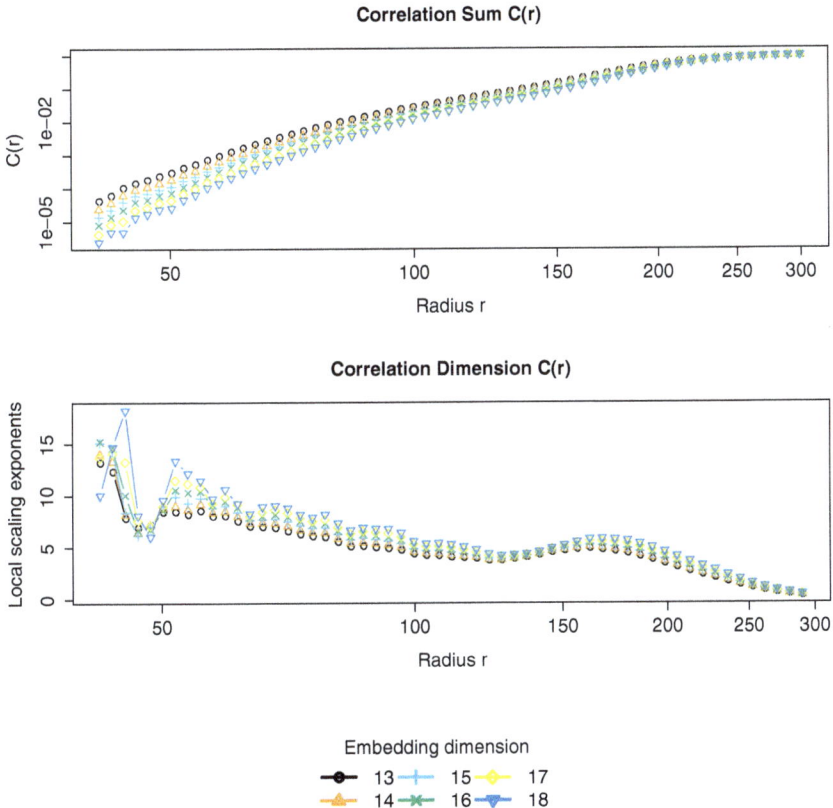

Fig. 5.10 Correlation sum of the "a05" recording after applying a noise reduction algorithm. The "staircase" behavior has disappeared

functions present in RHRV accept as inputs the standard parameters *HRVData*, *indexNonLinearAnalysis* and *doPlot*, as well as the following ones:

- *regressionRange*: two-dimensional vector specifying the range in which the scaling region appears and that, therefore, should be used for performing the regression.
- *useEmbeddings*: numeric vector specifying which embedding dimensions should be used to compute the correlation dimension.

Listing 5.4.9 illustrates the use of the *EstimateCorrDim* function. The estimation of the scaling exponent and all the computations performed are stored under the *hrv.apn$NonLinearAnalysis[[1]]$correlation* list (the subscript 1 is due to the use of the first nonlinear analysis structure). The correlation sums (as well as other information required for the estimations) are stored under the *computations* field, whereas the final estimation is stored under the *statistic* field. This nomenclature is maintained across the different nonlinear functions to store the calculations and the final estimations of the nonlinear analysis techniques.

R listing 5.4.9

```
HRVData <- EstimateCorrDim(hrv.apn,
                          regressionRange = c(120, 145),
                          useEmbeddings = 15:17,
                          doPlot = FALSE)
## Estimating the Correlation dimension
## Correlation dimension = 4.1462
```

5.4.5 Sample Entropy

From Eq. 5.4, it is clear that, once we have computed the correlation sum $C_q(m, r)$ for several embedding dimensions, it is possible to use these calculations to obtain an estimation of the sample entropy. The *CalculateSampleEntropy* routine calculates the $h_q(m, r)$ function using some previously computed correlation sums. Thus, this routine is very simple: it only takes as input parameters the *HRVData* structure, the *indexNonLinearAnalysis* (from which the correlation sums are obtained), and the Boolean argument *doPlot* (the *PlotSampleEntropy* function can also be used to obtain a plot of $h_q(m, r)$).

In a nonlinear system, $h_q(m, r)$ should converge toward the sample entropy H_q for sufficiently large m values. Hence, $h_q(m, r)$ should be approximately constant for all embedding dimensions larger than some m_{min} in some range of r. The *EstimateSampleEntropy* can be used for averaging $h_q(m, r)$ for some proper embedding dimensions (*useEmbeddings*) and radius (*regressionRange* parameter). If $h_q(m, r)$ is constant in a convincingly large range of scales, the resulting average can be interpreted as the sample entropy H_q of the system. Note that this parameters are the same as those used in *CalculateCorrDim*.

Listing 5.4.10 shows how to obtain an estimation of the sample entropy in RHRV. The $h_q(m, r)$ function and the H_q estimates for each dimension (dotted flat lines) are shown in Fig. 5.11. We may distinguish three regions in the $h_q(m, r)$ curve for a fixed embedding dimension m. The *macroscopic regime* appears for large radius, in which $h_q(m, r)$ starts decreasing toward 0. For smaller radius, we may find a plateau, which is a clear sign of determinism. Indeed, purely stochastic systems are characterized by the monotonically decreasing $h_q(m, r)$ function fulfilling $h_q(m, r) \approx -\ln r$. In deterministic systems, this behavior also appears for the smallest radius (the so-called *noise regime*), where noise predominates. Thus, Fig. 5.11 suggests that measurement noise is able to "hide" the deterministic dynamics of the RR intervals, which makes most part of the plateau inaccessible. Because of the small scaling region, we are not confident enough about our estimation to refer to it as the sample entropy of the system.

5.4 Chaotic Nonlinear Analysis with RHRV

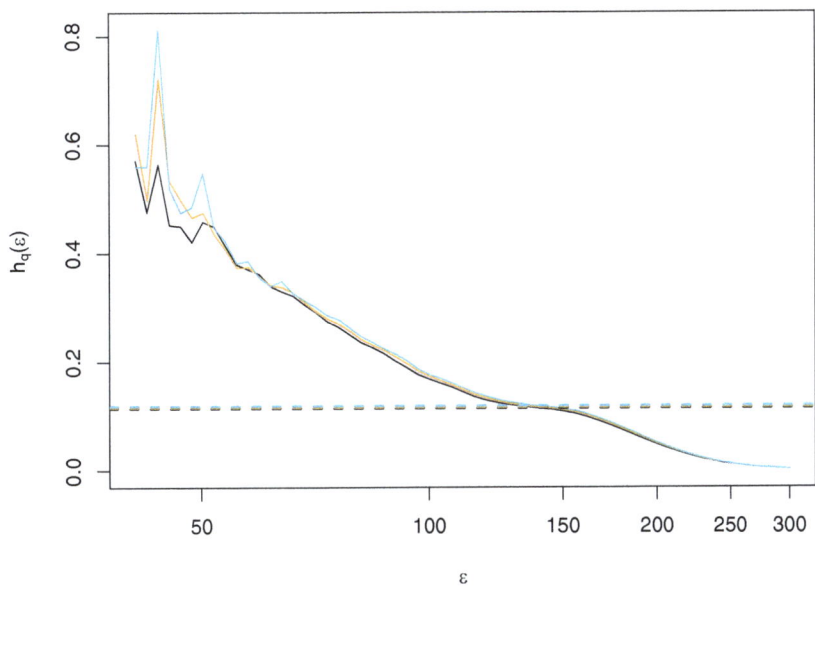

Fig. 5.11 Estimation of the sample entropy

R listing 5.4.10

```
hrv.apn <- CalculateSampleEntropy(hrv.apn, doPlot = FALSE)

## Computing the sample entropy of order 2

hrv.apn <- EstimateSampleEntropy(hrv.apn,
                                 regressionRange = c(130, 150),
                                 useEmbeddings = 15:17)

## Computing the sample entropy
## Sample entropy= 0.1169
```

5.4.6 Maximal Lyapunov Exponent

The procedure for obtaining an estimation of the maximal Lyapunov exponent is the same as for the fractal dimensions and the sample entropy. First, the $S(t)$ function (Eq. 5.6) is computed using the *CalculateMaxLyapunov* function. Since the computation of $S(t)$ can be computationally expensive, the *CalculateMaxLyapunov* implementation does not average the divergence of nearby trajectories over the whole phase space, but only in a few "reference points." To make computations more reliable, we require a minimum number of neighbors for Takens' vector to become a reference point. Thus, in addition to the standard parameters *HRVData*, *indexNonLinearAnalysis*, and *doPlot*, the *CalculateMaxLyapunov* function accepts as inputs the following:

- *minEmbeddingDim*, *maxEmbeddingDim*, *timeLag*, and *theilerWindow*, with the same meaning as in previous sections.
- *radius*: maximum distance in which the algorithm will search for nearby trajectories.
- *numberTimeSteps*: integer denoting the number of time steps in which the $S(t)$ function will be computed (default: 20 time steps).
- *minRefPoints*: number of reference points that the routine will try to use. The function stops when it finds *minRefPoints* reference points, saving computation time (default: 500). The minimum number of neighbors that Takens' vector must have in order to be considered a reference point can be specified using *minNeighs* (default: 5).

Once the $S(t)$ function has been computed, the maximal Lyapunov exponent is estimated through regression using the *EstimateMaxLyap* function, which takes the same input parameters as the others *EstimateX* functions (see Sects. 5.4.4 and 5.4.5): *regressionRange*, *useEmbeddings*, and *doPlot*.

R listing 5.4.11

```
hrv.apn <-
  CalculateMaxLyapunov(hrv.apn, minRefPoints = 1000,
                       numberTimeSteps = 25,
                       minEmbeddingDim = selectedED,
                       maxEmbeddingDim = selectedED + 2,
                       timeLag = selectedTL,
                       radius = 60,
                       theilerWindow = selectedTheiler,
                       doPlot = FALSE)

## Computing the divergence of the time series

hrv.apn <- EstimateMaxLyapunov(hrv.apn,
                               useEmbeddings = 13:15,
                               regressionRange = c(1, 8))

## Estimating the Maximum Lyapunov exponent
## Maximum Lyapunov exponent = 0.0558
```

5.4 Chaotic Nonlinear Analysis with RHRV

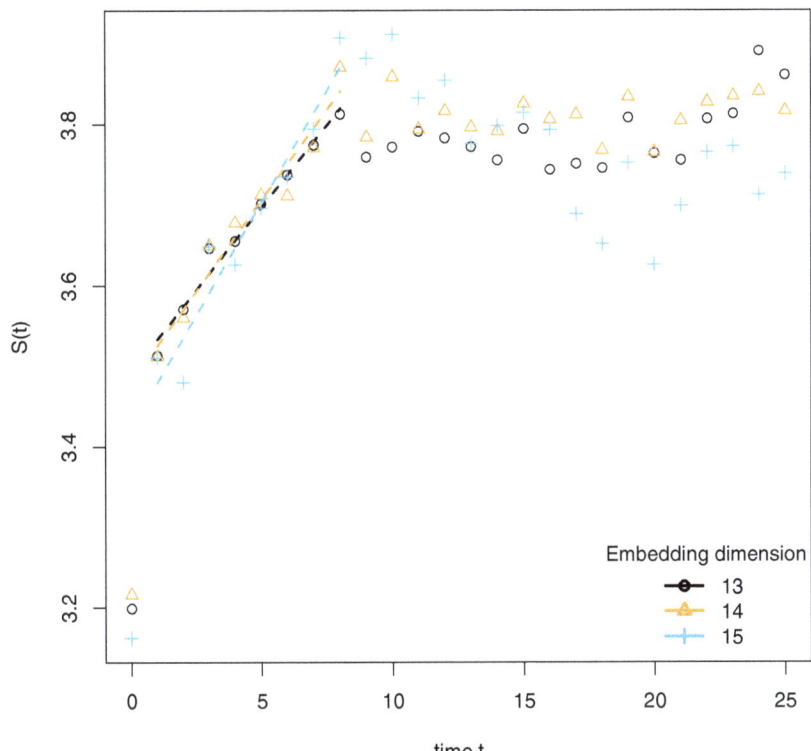

Fig. 5.12 Estimation of the maximal Lyapunov exponent

Listing 5.4.11 makes use of both *CalculateMaxLyapunov* and *EstimateMaxLyapunov* to obtain an estimation of the maximal Lyapunov exponent. The *radius* parameter was selected based on the correlation sum shown in Fig. 5.10. This decision is based on the interpretation of the correlation sum $C(r)$ as the average probability of finding a neighbor in a ball of radius r. Figure 5.12 shows the regressions performed to obtain the estimation of the maximal Lyapunov exponent. All the curves show a small region with more or less parallel slopes, which indicates a deterministic behavior. Also note that the $S(t)$ functions show oscillations due to the periodicity of the RR intervals. The resulting Lyapunov exponent is consistent with chaos phenomena ($0 < \lambda < \infty$). The *hrv.apn$NonLinearAnalysis[[1]]$lyapunov* list stores both the $S(t)$ functions (*computations* field) and the final estimation of the maximal Lyapunov exponent (*statistic* field).

5.4.7 RQA

The *RQA* function can be used for obtaining the recurrence plot of the RR intervals and quantifying the recurrences present in the signal. More specifically, the *RQA* function computes the statistics summarized in Table 5.1. In addition to the standard *HRVData*, *indexNonLinearAnalysis*, and *doPlot* parameters, the *RQA* function accepts as inputs the following:

- *embeddingDim* and *timeLag* with the same meaning as in previous sections.
- *radius*: maximum distance between two phase space points to be considered a recurrence.
- *lmin* and *vmin*: minimal lengths of diagonal lines and vertical lines (respectively) to be considered in the RQA computations (see Table 5.1). The default value is 2.
- In order to avoid border effects, the points near the border of the recurrence matrix are usually ignored when computing the RQA parameters. The *distanceToBorder* parameter specifies a minimum distance at which any point should be from the border to be included in the calculations.

Listing 5.4.12 computes all the RQA parameters and produces the RP shown in Fig. 5.13. Again, the radius was selected based on the correlation sum shown in Fig. 5.10. Since the recurrence matrix should be sparse, a radius of 73 can be selected, since the mean probability of finding a neighbor in this radius is approximately 0.005. As a consequence, the *REC* parameter will take a value close to 0.005 (see Listing 5.4.12). We have set *vmin=5*, and therefore, the trajectory must remain for five time steps in the same state to be considered as "trapped" in that state. We have also set *lmin=5* to compare the laminarity and the determinism parameters. All RQA parameters are stored in the *hrv.apn$NonLinearAnalysis[[1]]$rqa* list (with the name specified in Table 5.1) and can be accessed at any time.

The RP shown in Fig. 5.13 is quite homogeneous and provides no evidence of non-stationarity. The *TREND* parameter is close to zero and thus, it also supports stationarity. The diagonal lines in the RP indicate that the evolution of the process at different times is similar, which is a sign of determinism. However, the determinism parameter and the mean length of the diagonal lines of the RP are quite small ($DET = 13\%$ and $LmeanWithoutMain = 7.2$). This is probably due to the presence of noise which generates isolated recurrent points that "break" diagonal lines. As a further consequence, small diagonal lines predominate and the entropy is also small ($ENTR = 1.08$ Nats ≈ 1.56 bits). The laminarity and the mean length of vertical lines are even smaller than the statistics related to diagonal lines ($LAM = 2\%$ and $Vmean = 5.26$). Furthermore, the visual pattern is not apparent in this case. Hence, we may conclude that vertical structures are not common in this system.

5.4 Chaotic Nonlinear Analysis with RHRV

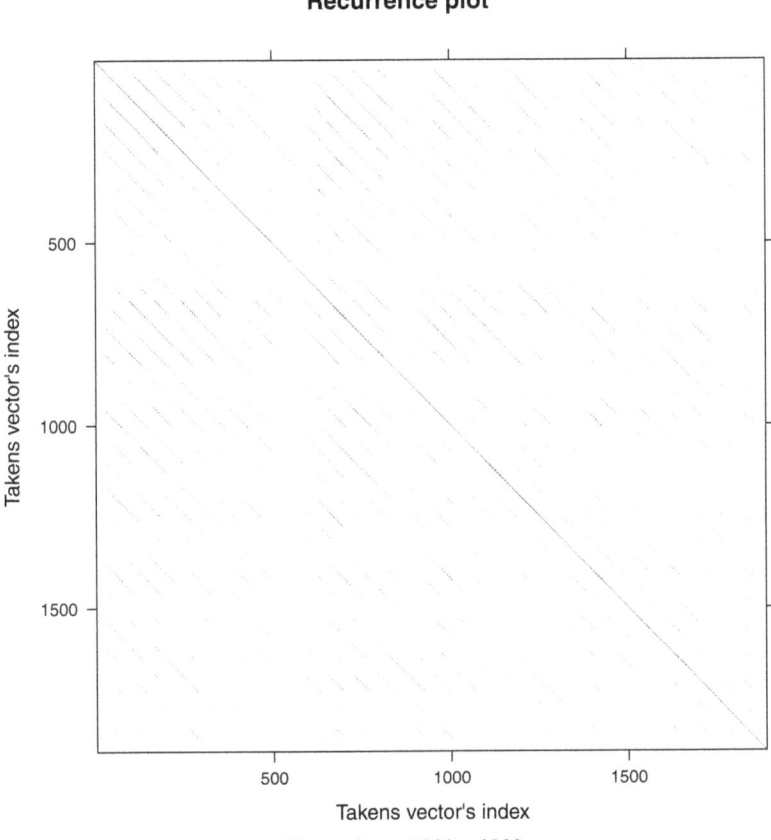

Fig. 5.13 RP of the "a05" recording. The harmonic behavior of the RR intervals is evident from this plot

The almost-periodic behavior of the RR series is also manifested in the RP, given that the distance between diagonal lines seems to be constant. However, it must be noted that Takens' vectors were built using the RR intervals, and thus, they are not equally spaced in time. We may refer to the behavior observed in the RP as almost-periodic only because the mean RR remains constant through the recording, and therefore the mean separation time between Takens' vectors is also constant. To study the almost-periodic behavior of the RR intervals, we may use the t-recurrence rate, since it may be interpreted as the probability that a state returns to its neighborhood after t steps. This is done in Listing 5.4.13. The resulting Fig. 5.14 clearly manifests the almost-periodic behavior of the RR intervals.

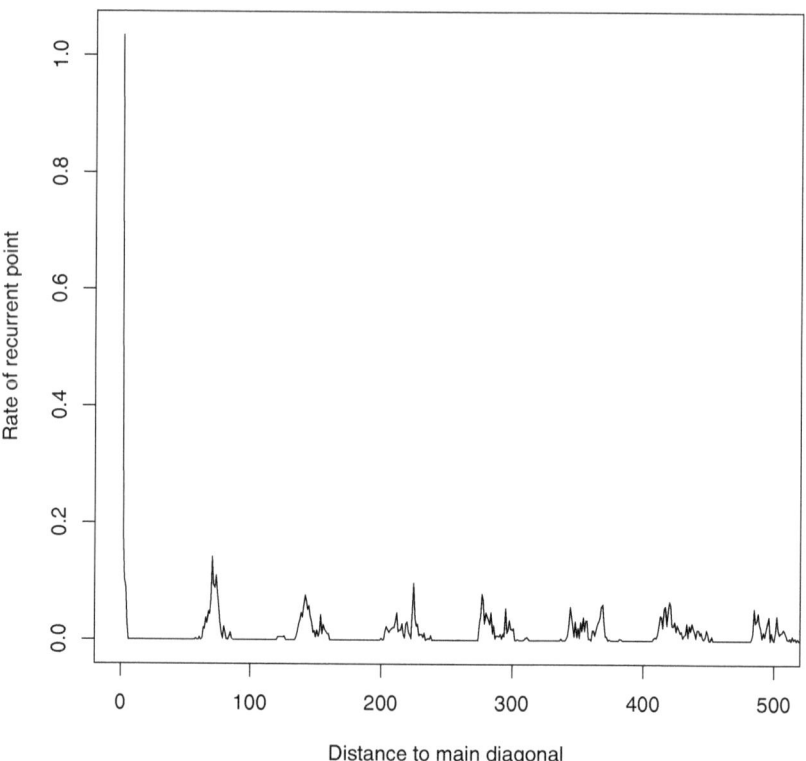

Fig. 5.14 t-recurrence rate of the signal. The harmonic behavior of the RP can be quantified using this measure

R listing 5.4.12

```
hrv.apn <-  RQA(hrv.apn,
                indexNonLinearAnalysis = 1,
                embeddingDim = selectedED,
                timeLag = selectedTL,
                lmin = 5, vmin = 5,
                radius = 73, doPlot = TRUE)

##  Performing Recurrence Quantification Analysis

# print the RQA parameters
str(hrv.apn$NonLinearAnalysis[[1]]$rqa[2:12])

   ## List of 11
   ##  $ REC     : num 0.00614
   ##  $ RATIO   : num 21.1
   ##  $ DET     : num 0.129
   ##  $ DIV     : num 0.0233
```

```
##  $ Lmax            : int 43
##  $ Lmean           : num 21.4
##  $ LmeanWithoutMain: num 7.2
##  $ ENTR            : num 1.08
##  $ TREND           : num -3.12e-06
##  $ LAM             : num 0.0228
##  $ Vmax            : int 7
```

R listing 5.4.13

```
recRate <- hrv.apn$NonLinearAnalysis[[1]]$rqa$recurrenceRate
plot(1:length(recRate), recRate, type = "l",
    main = "Recurrence Rate",
    xlab = "Distance to main diagonal",
    ylab = "Rate of recurrent point",
    xlim = c(0, 500))
```

5.4.8 Poincaré Plot

The *PoincarePlot* function computes both SD_1 and SD_2 parameters characterizing the Poincaré plot of $(RR_j, RR_{j+\tau})$. Two approaches can be used for computing these parameters: using Eqs. 5.7 and 5.8 or assuming a Gaussian distribution on $(RR_j, RR_{j+\tau})$ and computing its covariance matrix. In most cases, both approaches yield similar results. The most important parameters that can be used with the *PoincarePlot* are the following:

- *timeLag*: the τ parameter to construct the bidimensional variable $(RR_j, RR_{j+\tau})$.
- *confidenceEstimation*: if *TRUE*, the covariance matrix is used for computing the SD_1 and SD_2 parameters. If *FALSE*, Eqs. 5.7 and 5.8 are used.
- *confidence*: the confidence used for plotting the confidence ellipse. The default value is 0.95.

Listing 5.4.14 illustrates the use of the *PoincarePlot* function with $\tau = 1$ (the most typical time lag value used for this plot in the HRV literature). Note that both approaches yield almost the same results. The resulting Poincaré plots are shown in Fig. 5.15.

R listing 5.4.14

```
par(mfrow = c(1, 2))
hrv.apn <- PoincarePlot(hrv.apn,
                       confidence = 0.95,
                       timeLag = 1, doPlot = TRUE)
## Calculating SD1 and SD2 parameters
```

```
## Creating Poincare Plot with time lag = 1
## SD1 =   24.6036
## SD2 =   96.8547

hrv.apn <- CreateNonLinearAnalysis(hrv.apn)

## Creating non linear analysis
## Data has now  2  nonlinear analysis

hrv.apn <- PoincarePlot(hrv.apn,
                        confidenceEstimation = T,
                        confidence = 0.95, timeLag = 1,
                        doPlot = TRUE,
                        main = "Poincare plot \n(Covariance approach)")

## Calculating SD1 and SD2 parameters
## Creating Poincare Plot with time lag = 1
## SD1 =   24.6035
## SD2 =   96.8254

par(mfrow = c(1, 1))
```

5.5 Fractal Analysis with RHRV

In this section, we illustrate the use of the fractal analysis techniques presented in Sect. 5.3 for studying the RR intervals during spontaneous breathing (stored in the *HRVData* structure *hrv.norm*; see Listing 5.4.1). The objective of these techniques is the estimation of the Hurst exponent of the RR time series.

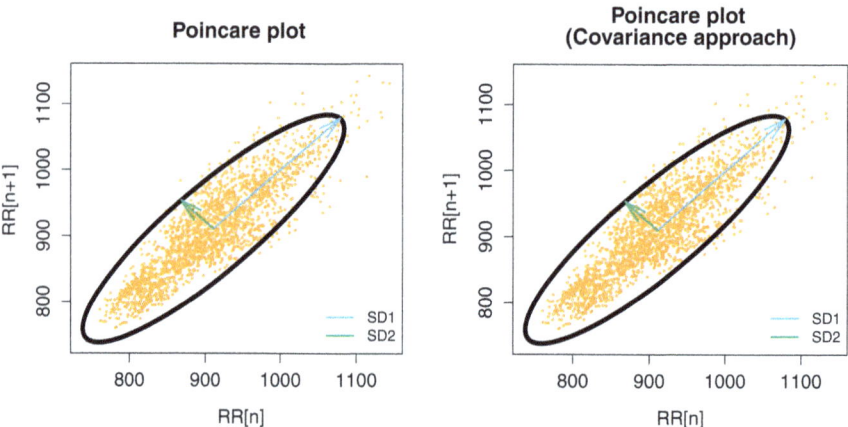

Fig. 5.15 Poincaré plot of the "a05" register with $\tau = 1$

5.5 Fractal Analysis with RHRV

RHRV stores each fractal analysis in the *NonLinearAnalysis* list (see Fig. 2.2). Thus, before performing a fractal analysis with RHRV, we should create the nonlinear analysis structure using the *CreateNonLinearAnalysis* function (see Listing 5.5.1).

5.5.1 Detrended Fluctuation Analysis

The DFA functions implemented in RHRV follow the same naming convention as the nonlinear analysis techniques: the *CalculateDFA* computes the fluctuation function for different window sizes and the *EstimateDFA* function estimates the scaling exponent using a least-squares fit. In addition to the standard *HRVData*, *indexNonLinearAnalysis*, and *doPlot* parameters, the *CalculateDFA* function accepts as inputs the following:

- *windowSizeRange*: range of values for the different window sizes (n in Sect. 5.3.1). Default: *c(10, 300)*
- *npoints*: the number of different window sizes to compute within the range specified by *windowSizeRange*

Listing 5.5.1 calculates and plots the fluctuation function using the *CalculateDFA* routine. The resulting plot is shown in Fig. 5.16. Note that the fluctuation function shows two different scaling behaviors (two different linear regions): one for $n < 20$ (that we shall denote with α_1) and another one for $n > 20$ (α_2). In non-pathological situations, α_1 is usually larger than α_2. This is probably due to the fact that in the shorter time scales, heart rate fluctuation is dominated by the smooth oscillation associated with respiration [30]. A $\hat{\alpha}_2 \approx 1$ was also reported for non-pathological data in [30].

R listing 5.5.1

```
hrv.norm <- CreateNonLinearAnalysis(hrv.norm)

## Creating non linear analysis
## Data has now  1  nonlinear analysis

hrv.norm <- CalculateDFA(hrv.norm,
                         windowSizeRange = c(5, 300),
                         npoints = 100, doPlot = TRUE)

## Performing Detrended Fluctuation Analysis
```

Listing 5.5.2 performs the estimation of both α_1 and α_2. Note that $\hat{\alpha}_1 > \hat{\alpha}_2$ and $\hat{\alpha}_2 \approx 1$, which suggest that the individual under study does not suffer from any severe pathology (despite suffering from apnea episodes) according to [30]. Note that there is uncertainty about the nature of the signal (either fGn or fBm) due to the proximity of the scaling exponents to 1.

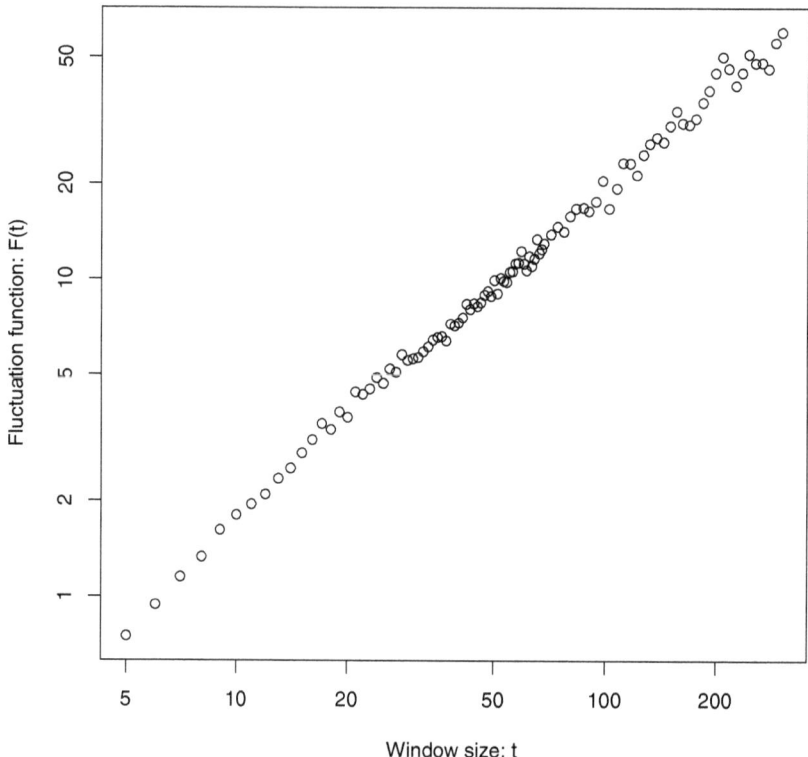

Fig. 5.16 DFA of the RR intervals from recording "a05" during normal breathing. It should be noted the existence of two different scaling exponents: one for $n < 20$ and another one for $n > 20$

R listing 5.5.2

```
hrv.norm <- EstimateDFA(hrv.norm,
                       regressionRange = c(5, 20),
                       doPlot = FALSE)

## Estimating Scaling exponent
## Scaling Exponent number 1 = 1.1795

hrv.norm <- EstimateDFA(hrv.norm,
                       regressionRange = c(20, 200),
                       doPlot = FALSE)

## Estimating Scaling exponent
## Scaling Exponent number 2 = 1.0196
```

The DFA computations are stored in the *dfa$computations* list in the *NonLinearAnalysis* structure. All the estimated exponents are stored under the *dfa$statistic* list. For example, the user can access the second scaling exponent estimate through *hrv.norm$NonLinearAnalysis[[1]]$dfa$statistic[[2]]$estimate*. With the index 1, we are accessing the first nonlinear analysis structure, whereas with the index 2, we are accessing the data from the second DFA estimate.

5.5.2 Power Spectral Analysis

The *EstimatePSDSlope* function estimates the spectral index of the RR intervals using some previously computed spectrogram. As usual, the *EstimatePSDSlope* function accepts as inputs the following: *HRVData*, *indexNonLinearAnalysis*, and *doPlot*. Additionally, the user can specify the following parameters:

- *indexFreqAnalysis*: an integer referencing the periodogram that will be used for estimating the spectral index. Note that the periodogram must be computed prior to the use of the *EstimatePSDSlope* function.
- *regressionRange*: range of frequencies in which the regression will be performed. Default is [1e-4,1e-2] Hz [31].

Listing 5.5.3 computes the spectral index and shows the fit of the PSD (see Fig. 5.17). Note that the resulting estimation yields a spectral index close to 1. Thus, the nature of the signal is uncertain, as the DFA suggested in Sect. 5.5.1.

R listing 5.5.3

```
hrv.norm <- SetVerbose(hrv.norm, F)
hrv.norm <- InterpolateNIHR(hrv.norm)
hrv.norm <- CreateFreqAnalysis(hrv.norm)
hrv.norm <- CalculatePSD(hrv.norm, indexFreqAnalysis = 1,
                    method = "pgram", doPlot = FALSE)
hrv.norm <- SetVerbose(hrv.norm, T)
hrv.norm <-
  EstimatePSDSlope(hrv.norm, indexFreqAnalysis = 1,
              indexNonLinearAnalysis = 1,
              regressionRange = c(5e-04, 3e-02))

## Calculating spectral index
## Plotting PSD
## Spectral index = 1.145
## H_fBm = 0.0725
```

The information related to the spectral index computations and the resulting estimation are stored in the *hrv.data$NonLinearAnalysis[[1]]$PSDSlope* list, as *computations* and *statistic*. Thus, the user can access the spectral index estimate through *hrv.norm$NonLinearAnalysis[[1]]$PSDSlope$statistic*.

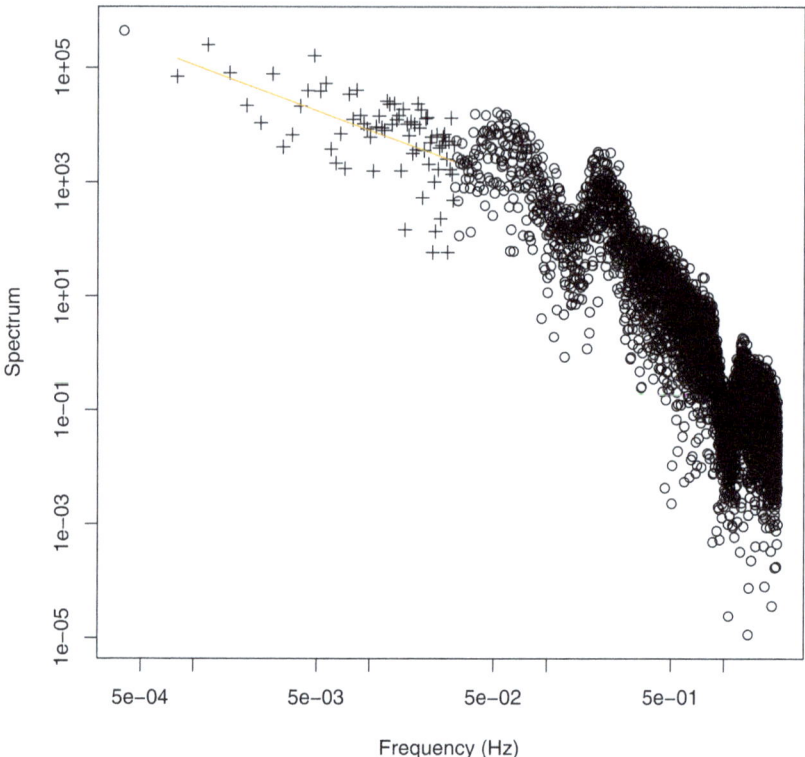

Fig. 5.17 Estimation of the spectral index

5.6 Nonlinear and Fractal Analysis of HRV Under Pathological Conditions

All the techniques presented in this chapter try to measure objectively the complexity of the RR time series. In the following paragraphs, we summarize the main findings from different studies related to the complex dynamics present in heart rhythm. We do not try to give an exhaustive list of the main articles in the field, but illustrate the major approaches to nonlinear/fractal HRV analysis. It must be noted that most of these findings agree with the rule that the more healthy a heart is, the more complex the HR is.

A reduction in the fractal properties of HR was reported in patients with congestive heart failure by applying DFA [30, 32]. In [32], the short-term fractal exponent was the strongest predictor of mortality among all the statistics tested. An interesting finding of the later study was that HRV statistics were strong predictors of mortality in patients with moderate congestive heart failure, but they fail to provide information for severe cases of this disease. Congestive heart failure patients also present a reduction in the LF power and lower $1/f$ slope, compared with controls [33].

After myocardial infarction or denervation of the heart, patients present reduced overall spectral power and a steeper slope in the power spectrum of the RR intervals. The power law parameter is valuable to predict death after myocardial infarction [34, 35]. In [35] other nonlinear statistics of HRV (SD_1 and SD_2 from the Poincaré plot, as well as the short-term fractal scaling exponent α_1), were associated with mortality after myocardial infarction.

Patients suffering from dilated cardiomyopathy present a reduced correlation dimension of its heartbeat intervals [36]. In this same study, the comparison of the correlation dimension during day and night suggested that the circadian rhythm was lost in patients suffering from dilated cardiomyopathy. Lower correlation dimension also precedes ventricular fibrillation [37].

A reduction in the sample entropy was found before the onset of atrial fibrillation [38] and before clinical signs of neonatal sepsis [39]. Children with cyanotic and acyanotic congenital heart disease display more complex HRV (measured by the approximate and sample entropies) than controls [40].

Although nonlinear techniques have been used extensively as markers of several diseases, its interpretation in terms of sympathetic and parasympathetic activity remains unclear. In a recent study, Bolea et al. demonstrated that parasympathetic blockade causes a significant decrease in all nonlinear statistics under study (correlation dimension, sample and approximate entropies), whereas the sympathetic blockade causes significant changes only in approximate entropy in standing position [41]. Sympathetic activation due to postural changes also caused a significant reduction in nonlinear statistics.

5.7 Some Final Remarks Regarding HRV Analysis with Chaotic and Fractal Techniques

It is not complicated to realize that if the heart rate control is a nonlinear chaotic system, it cannot be fractional Gaussian noise or fractional Brownian motion and vice versa. A system cannot be both completely deterministic (nonlinear chaotic) and completely stochastic (fGn or fBm). Furthermore, in a stochastic system, the Lyapunov exponent has an infinite value and the correlation dimension is not defined.

Therefore, which one is the right approach? Much has been written in the literature on this topic. Two good starting points for the reader interested in exploring this issue are [17, 42]. In this section, we will just offer some thoughts that reflect our view on the subject.

The RR series does have a broadband spectrum that could be compatible with the power law characterizing long memory processes. In addition, the RR series seems self-affine: rescaled fragments of the RR series have statistical properties similar to the entire series. These features support the use of fractal techniques in the study of HRV.

However, a completely stochastic model cannot explain some control mechanisms of HRV. Every doctor knows that the supply of norepinephrine results in a decrease of heart rate. Just the opposite effect can be obtained by supplying acetylcholine. Heart rate does not behave as a purely stochastic phenomenon, but there are "control variables" that influence it and that can be modified to try to drive the patient to a desired state. This supports the hypothesis that the HRV control system could be modeled through a set of deterministic equations. And there exists a family of deterministic phenomena that generates time series with a broadband spectrum (such as the RR series): those with nonlinear dynamics. However, it seems hard to believe that a complex biological system such as the heart can be completely described by a set of mathematical equations. Can these equations really predict when will I have an emotional reaction that increases my heart rate, or when am I going to start running?

Are there tests that should allow us to check whether the RR series is nonlinear chaotic, or fGn, or fBm? In theory, yes. We have covered them in this chapter. But these tests have to be applied to a finite, and typically smaller than we would like, number of observations. And these observations always have a measurement error. Hence, when we see that the spectrum of the RR series does not really behave as 1/f, but it has higher spectral power than it theoretically should for the high frequencies, we assume the measurement error to be the cause of these additional contributions to the high frequencies. Or when we see a strange behavior on some statistics derived from the chaos theory, we attribute this to the discretization error, and we apply a nonlinear noise reduction algorithm (which may be creating an artificial nonlinear structure in the data). The reality is that these tests, applied over finite and noisy datasets, seldom provide conclusive results.

It is curious to note that using techniques based on mutually exclusive assumptions on the same signal (RR series as a pure stochastic process versus RR series as a pure deterministic process), there have been found potential markers of several pathologies. A plausible explanation for this apparent paradox is that both hypotheses are partially correct.

We believe that HRV is the result of deterministic and stochastic phenomena that concur in the control of the myocardium. If this hypothesis is correct, trying to model HRV using a purely fractal or a purely deterministic model is a suboptimal approach, because it ignores part of the origin of the variability in heart rate. This could be the reason why fGn/fBm or nonlinear models are able to extract valuable statistics that correlate with multiple pathologies, although they are incapable of fully explaining HRV: each of these techniques could be capturing a different aspect of the underlying phenomena in HRV. However, when an analysis tool only captures part of the phenomena concurring in the system, we cannot achieve a full understanding of the system.

At present, we are trying to develop new algorithms for HRV analysis which assume that in the RR series, there are both deterministic and stochastic contributions. In the future, if these algorithms prove to be useful in the analysis of HRV, we will include them in RHRV.

References

1. H. Kantz, T. Schreiber, *Nonlinear Time Series Analysis*, vol. 7 (Cambridge University Press, Cambridge, 2004). [Online]. Available: https://doi.org/10.1017/CBO9780511755798
2. T. Schreiber, A. Schmitz, Surrogate time series. Physica D **142**(3), 346–382 (2000). [Online]. Available: https://doi.org/10.1016/S0167-2789(00)00043-9
3. J. Theiler, S. Eubank, A. Longtin, B. Galdrikian, J.D. Farmer, Testing for nonlinearity in time series: the method of surrogate data. Physica D **58**(1), 77–94 (1992). [Online]. Available: https://doi.org/10.1016/0167-2789(92)90102-S
4. R. Mañé, On the dimension of the compact invariant sets of certain non-linear maps, in *Dynamical Systems and Turbulence, Warwick 1980* (Springer, Berlin, 1981), pp. 230–242. [Online]. Available: https://doi.org/10.1007/BFb0091916
5. F. Takens, *Detecting Strange Attractors in Turbulence*, vol. 898 (Springer, Berlin, 1981). [Online]. Available: https://doi.org/10.1007/BFb0091903
6. H.D.I. Abarbanel, C. Institute for nonlinear science (San Diego), *Analysis of Observed Chaotic Data*, ser. Institute for nonlinear science (Springer, Berlin, 1996). [Online]. Available: https://doi.org/10.1007/978-1-4612-0763-4
7. L. Cao, Practical method for determining the minimum embedding dimension of a scalar time series. Physica D **110**(1), 43–50 (1997). [Online]. Available: https://doi.org/10.1016/S0167-2789(97)00118-8
8. A.M. Fraser, H.L. Swinney, Independent coordinates for strange attractors from mutual information. Phys. Rev. A **33**(2), 1134–1140 (1986). [Online]. Available: https://doi.org/10.1103/PhysRevA.33.1134
9. P. Grassberger, I. Procaccia, Characterization of strange attractors. Phys. Rev. Lett. **50**(5), 346–349 (1983). [Online]. Available: https://doi.org/10.1103/PhysRevLett.50.346
10. S.M. Pincus, Approximate entropy as a measure of system complexity. Proc. Natl. Acad. Sci. **88**(6), 2297–2301 (1991). [Online]. Available: https://doi.org/10.1073/pnas.88.6.2297
11. J.S. Richman, J.R. Moorman, Physiological time-series analysis using approximate entropy and sample entropy. Am. J. Physiol. Heart Circ. Physiol. **278**(6), H2039–H2049 (2000). [Online]. Available: https://doi.org/10.1152/ajpheart.2000.278.6.H2039
12. J.P. Zbilut, C.L. Webber, Embeddings and delays as derived from quantification of recurrence plots. Phys. Lett. A **171**(3), 199–203 (1992). [Online]. Available: https://doi.org/10.1016/0375-9601(92)90426-M
13. C.L. Webber, J.P. Zbilut, Dynamical assessment of physiological systems and states using recurrence plot strategies. J. Appl. Physiol. **76**(2), 965–973 (1994). [Online]. Available: https://doi.org/10.1152/jappl.1994.76.2.965
14. C.L. Webber Jr, J.P. Zbilut, Recurrence quantification analysis of nonlinear dynamical systems, in *Tutorials in Contemporary Nonlinear Methods for the Behavioral Sciences* (2005), pp. 26–94
15. M. Brennan, M. Palaniswami, P. Kamen, Do existing measures of poincare plot geometry reflect nonlinear features of heart rate variability? IEEE Trans. Biomed. Eng. **48**(11), 1342–1347 (2001). [Online]. Available: https://doi.org/10.1109/10.959330
16. B.B. Mandelbrot, How long is the coast of britain. Science **156**(3775), 636–638 (1967). [Online]. Available: https://doi.org/10.1126/science.156.3775.636
17. A. Eke, P. Herman, J. Bassingthwaighte, G. Raymond, D. Percival, M. Cannon, I. Balla, C. Ikrényi, Physiological time series: distinguishing fractal noises from motions. Pflügers Archiv **439**(4), 403–415 (2000). [Online]. Available: https://doi.org/10.1007/s004249900135

18. R.T. Baillie, A.A. Cecen, C. Erkal, Normal heartbeat series are nonchaotic, nonlinear, and multifractal: new evidence from semiparametric and parametric tests. Chaos Interdiscip. J. Nonlin. Sci. **19**(2), 028503 (2009). [Online]. Available: https://doi.org/10.1063/1.3152006
19. R. Sassi, M.G. Signorini, S. Cerutti, Multifractality and heart rate variability. Chaos Interdiscip. J. Nonlin. Sci. **19**(2), 028507 (2009). [Online]. Available: https://doi.org/10.1063/1.3152223
20. B.B. Mandelbrot, Self-affine fractals and fractal dimension. Phys. Scr. **32**(4), 257–260 (1985). [Online]. Available: https://doi.org/10.1088/0031-8949/32/4/001
21. C.-K. Peng, S.V. Buldyrev, S. Havlin, M. Simons, H.E. Stanley, A.L. Goldberger, Mosaic organization of DNA nucleotides. Phys. Rev. E **49**(2), 1685–1689 (1994). [Online]. Available: https://doi.org/10.1103/physreve.49.1685
22. P.F. Fougere, On the accuracy of spectrum analysis of red noise processes using maximum entropy and periodogram methods: simulation studies and application to geophysical data. J. Geophys. Res.: Space Phys. (1978–2012) **90**(A5), 4355–4366 (1985). [Online]. Available: https://doi.org/10.1007/978-94-009-3961-5_8
23. V.K. Somers, D.P. White, R. Amin, W.T. Abraham, F. Costa, A. Culebras, S. Daniels, J.S. Floras, C.E. Hunt, L.J. Olson, Sleep apnea and cardiovascular disease: An american heart association/american college of cardiology foundation scientific statement from the american heart association council for high blood pressure research professional education committee, council on clinical cardiology, stroke council, and council on cardiovascular nursing in collaboration with the national heart, lung, and blood institute national center on sleep disorders research (national institutes of health). J. Am. Coll. Cardiol. **52**(8), 686–717 (2008). [Online]. Available: https://doi.org/10.1161/CIRCULATIONAHA.107.189375
24. T. Penzel, G. Moody, R. Mark, A. Goldberger, J. Peter, The Apnea-ECG database, in *Computers in Cardiology 2000* (IEEE, Piscataway, 2000), pp. 255–258. [Online]. Available: https://doi.org/10.1109/CIC.2000.898505
25. The Apnea-ECG database. Last accessed 23 Apr 2024. [Online]. Available: http://www.physionet.org/physiobank/database/apnea-ecg/
26. C. Guilleminault, R. Winkle, S. Connolly, K. Melvin, and A. Tilkian, Cyclical variation of the heart rate in sleep apnoea syndrome: mechanisms, and usefulness of 24 h electrocardiography as a screening technique. Lancet **323**(8369), 126–131 (1984). [Online]. Available: https://doi.org/10.1016/s0140-6736(84)90062-x
27. M. Kobayashi, T. Musha, 1/f fluctuation of heartbeat period. IEEE Trans. Biomed. Eng. 456–457 (1982). [Online]. Available: https://doi.org/10.1109/TBME.1982.324972
28. Task Force of the European Society of Cardiology and the North American Society of Pacing and Electrophysiology, Heart rate variability: Standards of measurement, physiological interpretation and clinical use. Eur. Heart J. **17**, 354–381 (1996). [Online]. Available: https://doi.org/10.1093/oxfordjournals.eurheartj.a014868
29. M. Möller, W. Lange, F. Mitschke, N. Abraham, U. Hübner, Errors from digitizing and noise in estimating attractor dimensions. Phys. Lett. A **138**(4), 176–182 (1989). [Online]. Available: https://doi.org/10.1016/0375-9601(89)90023-6
30. C.-K. Peng, S. Havlin, H.E. Stanley, A.L. Goldberger, Quantification of scaling exponents and crossover phenomena in nonstationary heartbeat time series. Chaos Interdiscip. J. Nonlin. Sci. **5**(1), 82–87 (1995). [Online]. Available: https://doi.org/10.1063/1.166141
31. A. Voss, S. Schulz, R. Schroeder, M. Baumert, P. Caminal, Methods derived from nonlinear dynamics for analysing heart rate variability. Phil. Trans. R. Soc. A **367**(1887), 277–296 (2009). [Online]. Available: https://doi.org/10.1098/rsta.2008.0232
32. T.H. Mäkikallio, H.V. Huikuri, U. Hintze, J. Videbæk, R.D. Mitrani, A. Castellanos, R.J. Myerburg, M. Møller, Fractal analysis and time-and frequency-domain measures of heart rate variability as predictors of mortality in patients with heart failure. Am. J. Cardiol. **87**(2), 178–182 (2001). [Online]. Available: https://doi.org/10.1016/s0002-9149(00)01312-6
33. S. Guzzetti, S. Mezzetti, R. Magatelli, A. Porta, G. De Angelis, G. Rovelli, A. Malliani, Linear and non-linear 24 h heart rate variability in chronic heart failure. Auton. Neurosci. **86**(1), 114–119 (2000). [Online]. Available: https://doi.org/10.1016/S1566-0702(00)00239-3

References

34. J.T. Bigger, R.C. Steinman, L.M. Rolnitzky, J.L. Fleiss, P. Albrecht, R.J. Cohen, Power law behavior of RR-interval variability in healthy middle-aged persons, patients with recent acute myocardial infarction, and patients with heart transplants. Circulation **93**(12), 2142–2151 (1996). [Online]. Available: https://doi.org/10.1161/01.cir.93.12.2142
35. P.K. Stein, P.P. Domitrovich, H.V. Huikuri, R.E. Kleiger, Traditional and nonlinear heart rate variability are each independently associated with mortality after myocardial infarction. J. Cardiovasc. Electrophysiol. **16**(1), 13–20 (2005). [Online]. Available: https://doi.org/10.1046/j.1540-8167.2005.04358.x
36. R. Carvajal, N. Wessel, M. Vallverdú, P. Caminal, A. Voss, Correlation dimension analysis of heart rate variability in patients with dilated cardiomyopathy. Comput. Methods Programs Biomed. **78**(2), 133–140 (2005). [Online]. Available: https://doi.org/10.1016/j.cmpb.2005.01.004
37. J.E. Skinner, C.M. Pratt, T. Vybiral, A reduction in the correlation dimension of heartbeat intervals precedes imminent ventricular fibrillation in human subjects. Am. Heart J. **125**(3), 731–743 (1993). [Online]. Available: https://doi.org/10.1016/0002-8703(93)90165-6
38. V. Tuzcu, S. Nas, T. Börklü, A. Ugur, Decrease in the heart rate complexity prior to the onset of atrial fibrillation. Europace **8**(6), 398–402 (2006). [Online]. Available: https://doi.org/10.1093/europace/eul031
39. D.E. Lake, J.S. Richman, M.P. Griffin, J.R. Moorman, Sample entropy analysis of neonatal heart rate variability. Am. J. Physiol. Regul. Intgr. Comp. Physiol. **283**(3), R789–R797 (2002). [Online]. Available: https://doi.org/10.1152/ajpregu.00069.2002
40. F. Aletti, M. Ferrario, T. Almas de Jesus, R. Stirbulov, A. Borghi Silva, S. Cerutti, L. Malosa Sampaio, Heart rate variability in children with cyanotic and acyanotic congenital heart disease: analysis by spectral and non linear indices, in *Annual International Conference of the IEEE, Engineering in Medicine and Biology Society (EMBC)* (IEEE, Piscataway, 2012), pp. 4189–4192. [Online]. Available: https://doi.org/10.1109/EMBC.2012.6346890
41. J. Bolea, E. Pueyo, P. Laguna, R. Bailón, Non-linear HRV indices under autonomic nervous system blockade, in *36th Annual International Conference of the IEEE, Engineering in Medicine and Biology Society (EMBC)* (IEEE, Piscataway, 2014), pp. 3252–3255. [Online]. Available: https://doi.org/10.1109/EMBC.2014.6944316
42. L. Glass, Introduction to controversial topics in nonlinear science: is the normal heart rate chaotic? Chaos Interdiscip. J. Nonlin. Sci. **19**(2), 028501 (2009). [Online]. Available: https://doi.org/10.1063/1.3156832

Chapter 6
Comparing HRV Across Different Segments of a Recording

Abstract Intervals of physiological interest within the heartbeat time series (such as apnea and ischemia) may be marked making use of tags in so-called episodes. Episodic information can be incorporated into RHRV from external files or can be added using functions included in the package. This information can be useful to compare sections of HRV records both visually and numerically.

6.1 Episodes and Physiological Events

It has been stated in the previous chapters of this book that the study of the HRV of the whole ECG recording can be employed to extract useful information about the general state of the subject. However, sometimes it can be interesting to analyze only several parts (episodes) of the heartbeat time series, since they can provide findings about what is happening at a specific moment.

Episodes can be defined as different types of events or intervals with physiological interest in the same heart rate recording, which occur at a determined moment and have a defined ending. In this way, episodes are characterized by both initial time and duration.

These important events in the heart rate recordings may come from different sources. In many occasions, episodes are identified by specialists, who label these physiological events, and the corresponding annotation files are generated from the clinicians' indications. This is the case, for example, of several databases that can be freely downloaded from PhysioNet [1], such as the MIT-BIH Atrial Fibrillation Database [2], that includes, for each EGC recording, rhythm annotations of atrial fibrillation, atrial flutter, and junctional rhythm. Other examples of different episodes can be found in the MIT-BIH Polysomnographic Database [3], which contains annotations corresponding to sleep stages, including when the subject is awake, as well as the different sleep cycles and the REM (rapid eye movement) stage, and apnea episodes, with information about respiratory events.

Software dedicated to the automated annotation of episodes can also be found. One example is VARVI, already cited in Chapter 2 [4]. Another possibility is Signal Labeler app, included in the Matlab computing environment [5].

From the previous paragraphs, it can be deduced the interest of analyzing episodes of different and varied types. The task can be easily performed employing the RHRV software package, which allows the user to study specific events in terms of HRV parameters. In Sect. 5.4, the *ExtractTimeSegment* function was presented, which allows users to extract segments of interest from recordings. In this chapter, we will describe more tools within RHRV to mark and analyze those segments of interest in HRV records, by adding information to these records. This is a flexible and powerful approach that will be described exhaustively in the following sections.

6.2 Using Episodes in RHRV

In Chap. 2, the *HRVData* structure which RHRV uses to store all the information related to a HR record was presented. As it can be seen in Fig. 2.2, RHRV can include one or more episodes, each one comprising four fields:

- *InitTime*: beginning of the episode from the start of the HR record (in seconds).
- *Type*: a string or *tag* used to label the episode.
- *Duration*: length of the episode (in seconds).
- *Value*: a value associated with the episode. This field is not used by any of the functions in RHRV, but some users may find it useful. For example, *Type* can be used to store the name of a drug administered to a patient, while *Value* can store the different dosages.

Episodes are the basic tool in RHRV to mark sections of the signal for plotting, extracting segments, or estimating certain features. In an HR record, episodes labels or tags can be reused to indicate repetitive events. Besides, episodes can overlap and there is no limit on the number of episodes on a record. Episodes are stored internally as a *dataframe* with name *Episodes* inside the *RHRVData* structure (see Fig. 2.2).

When dealing with episodes in RHRV, depending on the function, users have two possibilities to refer to them:

- By their labels or tags
- By their indices

In many cases, these two possibilities can be used simultaneously. However, there are functionalities where specifying multiple tags or indices does not make sense, and in these functions only a tag can be used. Readers are referred to the help pages for the distinct functions in *RHRV*.

Information on tags and indices can be shown by using the *ListEpisodes* function. For instance, Listing 6.2.1 shows the possible output obtained when using this function on data that contain three episodes named *"Before"*, *"During"*, and *"After"*.

R listing 6.2.1

```
## Index        Tag   InitTime   Duration   Value
## 1         Before        700        900       0
## 2         During       2000       2000       0
## 3          After       5000        600       0
```

6.2.1 Managing Episodes in a HR Record

There are two ways of adding episodic information to a HR record: loading it from an external file or creating it from scratch or from other episodes. For loading episodes, RHRV offers two functions: *LoadEpisodesAscii* and *LoadApneaWFDB*.

LoadEpisodesAscii accepts, among its arguments, a file that must be in ASCII format and must include one line per episode. The separator between columns is "white space," which is one or more spaces, tabs, or newlines. For example, Listing 6.2.2 shows the result of loading a file named *apnea_ascii.txt* (this file can be found in the GitHub repository that contains the code used in this book [6]) containing the following lines:

```
Init_Time    Resp_Events    Durat    SaO2
00:01:30     GEN_HYPO        12.0    82.9
00:04:00     OBS_APNEA        6.0    81.0
00:06:00     GEN_HYPO        12.0    82.9
```

The first line of file *apnea_ascii.txt* is skipped by default. This behavior can be changed setting *header=FALSE* when calling *LoadEpisodesAscii*.

R listing 6.2.2

```
library(RHRV)
data(HRVData)
hrv.data <- HRVData
hrv.data <- BuildNIHR(hrv.data)
hrv.data <- LoadEpisodesAscii(hrv.data,
  "./sampleData/apnea_ascii.txt")
ListEpisodes(hrv.data)
```

```
## Index           Tag  InitTime  Duration   Value
## 1          GEN_HYPO        90        12    82.9
## 2         OBS_APNEA       240         6      81
## 3          GEN_HYPO       360        12    82.9
```

LoadApneaWFDB is a function specially tailored to read annotations stored in WFDB format, often used by researchers in this field. It reads apnea episodes from .apn files present in some of the PhysioBank databases [7] available in PhysioNet [1]. These files are binary annotation files, each recording indicating the presence or absence of apnea at that time.

In Listing 6.2.3, an example of how to load beats and apnea annotations from a record belonging to "The Apnea ECG Database" [8] is included. This script downloads the files a01.hea, a01.qrs, and a01.apn; thus, an active Internet connection is necessary for it to work.

R listing 6.2.3

```
dirorig <-
  "http://www.physionet.org/physiobank/database/apnea-ecg/"
files <- c("a01.hea", "a01.apn", "a01.qrs")
filesorig <- paste(dirorig, files, sep = "")
for (i in 1:length(files))
  download.file(filesorig[i], files[i])
library(RHRV)
hrv.data <- CreateHRVData()
hrv.data <- LoadBeatWFDB(hrv.data, "a01")
hrv.data <- LoadApneaWFDB(hrv.data, "a01")
```

Listing 6.2.4 offers an alternative to Listing 6.2.3. Beats and tags are loaded using functions *LoadBeatAscii* and *LoadEpisodesAscii*, respectively (the files can be found in the GitHub repository that contains the code used in this book [6]).

R listing 6.2.4

```
library(RHRV)
hrv.data <- CreateHRVData()
hrv.data <- LoadBeatAscii(hrv.data,
   "./sampleData/a01.beats.txt")
hrv.data <- LoadEpisodesAscii(hrv.data,
   "./sampleData/a01.tags.txt")
```

New episodes can be created using *AddEpisodes*. The arguments of this function include the following:

- *InitTimes*: vector with the beginning time of each episode
- *Durations*: vector with each episode duration
- *Tags*: vector containing the tags of each episode
- *Values*: vector containing the value of each episode

For example, Listing 6.2.5 adds two new episodes to the data loaded in Listing 6.2.3.

R listing 6.2.5

```
hrv.data <- AddEpisodes(hrv.data,
   InitTimes = c(5000, 15000), Durations = c(2000, 2000),
   Tags = c("Tag1", "Tag2"), Values = c(0, 0))
ListEpisodes(hrv.data)
##    Index      Tag    InitTime    Duration    Value
##        1    APNEA         750       12840        0
```

6.2 Using Episodes in RHRV

```
##     2        Tag1    5000    2000    0
##     3       APNEA   13830    7260    0
##     4        Tag2   15000    2000    0
##     5       APNEA   21210    8070    0
```

Creating new episodes from previous ones is also possible by means of the *GenerateEpisodes* function. The preexisting episodes to be used can be selected by their tags. The following arguments allow users to specify how to calculate starting and ending times of new episodes:

- *NewBegFrom*
- *NewEndFrom*
- *DispBeg*
- *DispEnd*

For example, Listing 6.2.6 will select all episodes tagged *"APNEA"* and, for each one, it will create a new one tagged *"Pre-APNEA"* starting 70 s before its beginning and ending 10 s before its beginning, that is, a minute long episode ending just 10 s before the old one. If displacements should be calculated from the end of the previous episodes, users can use "End" instead of "Beg" in arguments *NewBegFrom* and *NewEndFrom*.

R listing 6.2.6

```
hrv.data <- GenerateEpisodes(hrv.data, NewBegFrom = "Beg",
    NewEndFrom = "Beg", DispBeg = -70, DispEnd = -10,
    OldTag = "APNEA", NewTag = "Pre-APNEA")
ListEpisodes(hrv.data)
##   Index         Tag   InitTime   Duration   Value
##     1      Pre-APNEA       680         60       0
##     2          APNEA       750      12840       0
##     3           Tag1      5000       2000       0
##     4      Pre-APNEA     13760         60       0
##     5          APNEA     13830       7260       0
##     6           Tag2     15000       2000       0
##     7      Pre-APNEA     21140         60       0
##     8          APNEA     21210       8070       0
```

There are two other functions that can be very useful when dealing with episodes: *RemoveEpisodes* and *ModifyEpisodes*. *RemoveEpisodes* deletes episodes from the HR record. The episodes to remove are selected by their tags or their indices, as can be seen in Listing 6.2.7.

R listing 6.2.7

```
hrv.data <- RemoveEpisodes(hrv.data, Tags = "APNEA",
    Indexes = c(1, 6))
ListEpisodes(hrv.data)
```

```
## Index             Tag    InitTime  Duration  Value
##    1             Tag1        5000      2000      0
##    2         Pre-APNEA      13760        60      0
##    3         Pre-APNEA      21140        60      0
```

With *ModifyEpisodes* all the fields within episodes can be changed. As with *RemoveEpisodes*, tags or indices (or both) can be used to specify which episodes to modify (see Listing 6.2.8). After the modification has been made, possible duplicated episodes are removed and they are reordered by increasing *InitTimes*.

R listing 6.2.8

```
hrv.data <- ModifyEpisodes(hrv.data, Tags = "Tag1",
   Indexes = 3, NewDurations = c(300, 400),
   NewValues = 8.5)
ListEpisodes(hrv.data)

## Index             Tag    InitTime  Duration  Value
##    1             Tag1        5000       300    8.5
##    2         Pre-APNEA      13760        60      0
##    3         Pre-APNEA      21140       400    8.5
```

There are two functions that, while not directly managing episodes, are related with specifying temporal segments of records and then should be commented here. These functions are *LoadBeatAscii* and *ExtractTimeSegment*. Both functions can use two arguments *starttime* and *endtime* (both in seconds) which specify the segment of the record we are dealing with.

- In *LoadBeatAscii*, *starttime* and *endtime* allow users to select the fragment of the beats sequence before loading it into the data structure.
- *ExtractTimeSegment* offers users the possibility of extracting a temporal subset of a previously loaded record, creating a new data structure, as explained in Sect. 5.4.

6.3 Using Episodes in Plots

RHRV includes functions to plot HR records and parameters estimated from these records. The most basic plotting functions are *PlotNIHR* and *PlotHR*, which represent the non-interpolated HR and the interpolated HR versus time, respectively. Both these functions allow users to specify which episodes to include by making use of the arguments *Tags* and *Indexes*. For example, let us suppose we have a data structure containing the episodes shown in Listing 6.3.1. Figures 6.1 and 6.2 could be created specifying the episodes to plot by making use of the arguments *Tags* = "all" and *Indexes* = c(1, 2), respectively.

6.3 Using Episodes in Plots

R listing 6.3.1

```
## Index           Tag InitTime Duration Value
##     1     Pre-apnea     3920     2000     0
##     2         Apnea     6270     2520     0
##     3    Post-apnea     9140     1840     0
```

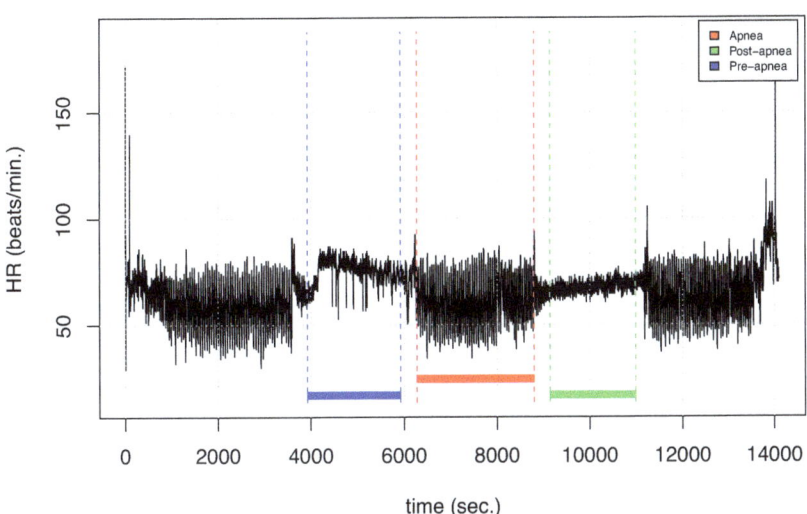

Fig. 6.1 Plot obtained using Tags ="all" in PlotNIHR

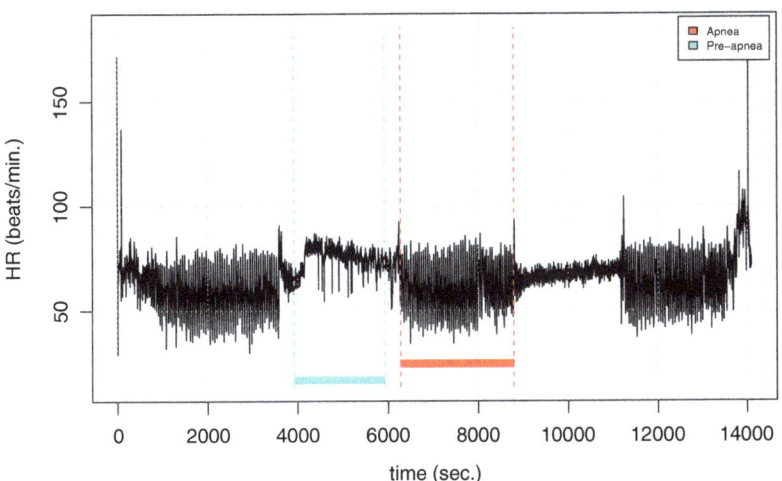

Fig. 6.2 Plot obtained using Indexes = c(1, 2) in PlotHR

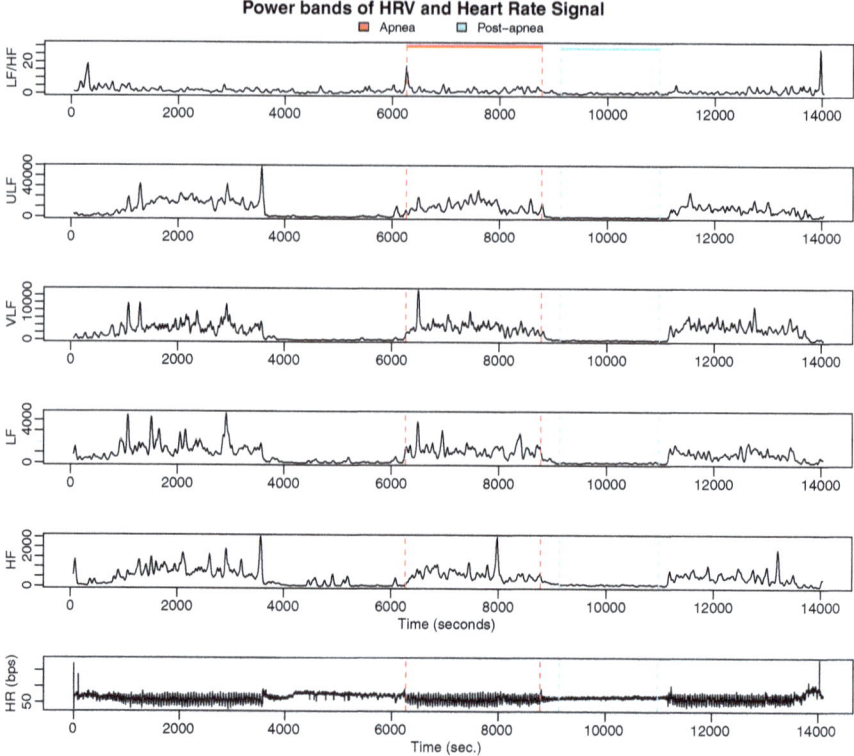

Fig. 6.3 Plot obtained using Tags = c("Apnea," "Post-apnea") in PlotPowerBand

Tags and *Indexes* can also be used in *PlotPowerBand*. For instance, Fig. 6.3 is obtained as a result of using *PlotPowerBand* on the same data as before, with argument *Tags = c("Apnea," "Post-apnea")*. Other similar functions that use episodes in the same way are *PlotSinglePowerBand* and *OverplotEpisodes*. Interested readers are suggested to consult the help pages of these functions for examples and further information.

Episodes can also be included when plotting spectrograms, as can be seen in Fig. 6.4, where *Indexes = c(2, 3)* was passed as argument to the function *PlotSpectrogram*.

6.4 Making Use of Episodes in HRV Analysis

Until this point, episodic information was used in plot functions to mark sections in figures generated by *RHRV*. However, researchers often want to compare parameters extracted from the HR record in different sections of the signal. This

6.4 Making Use of Episodes in HRV Analysis

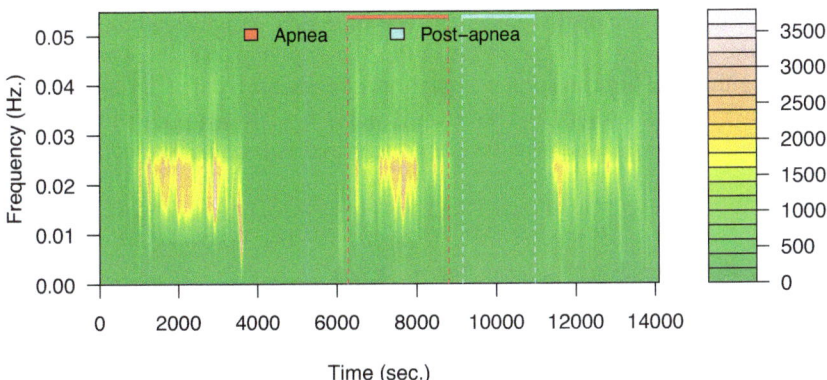

Fig. 6.4 Plot obtained using Indexes = c(2, 3) in PlotSpectrogram

can be easily achieved making use of several functions provided by *RHRV*, namely *SplitHRbyEpisodes*, *SplitPowerBandByEpisodes*, *AnalyzeHRbyEpisodes*, and *AnalyzePowerBandsByEpisodes*.

There is a basic difference between these functions and the plotting ones presented in Sect. 6.3. Plotting functions make use of arguments *Tags* and *Indexes* to select which sections to mark in the plots, allowing users to specify multiple episodes and tags simultaneously. However, functions presented in this section use the argument *Tag* instead of *Tags*. While *Tags* can be a list of strings, *Tag* can only be a single string, thus restricting the selection of the sections to gather information from. This may seem to be a drawback, but it is easily solved by making use of the functions provided to manage episodes that were presented in Sect. 6.2.1.

SplitHRbyEpisodes divides heart rate data into two parts: one inside episodes corresponding to the argument *Tag* and the other with the rest of the heart rate, as it can be seen in Listing 6.4.1. Users must be aware that segments of heart rate may be concatenated and that this can cause serious discontinuities in the resulting segments.

R listing 6.4.1

```
SplitData <- SplitHRbyEpisodes(hrv.data, Tag = "Apnea")
str(SplitData)

   ## List of 2
   ##  $ InEpisodes : num [1:10080] 69 69.7 70.4 71.2 72 ...
   ##  $ OutEpisodes: num [1:46314] 171 171 171 165 135 ...
```

SplitPowerBandByEpisodes is very similar to *SplitHRbyEpisodes*. *SplitPowerBandByEpisodes* returns a list including two sublists: *InEpisodes* and *OutEpisodes*, and both sublists will include ULF, LF, LF, and HF bands.

If users want to split and apply some procedure to heart rate data on different sections marked by episodes, they may perform this by using *SplitHRbyEpisodes* and processing the obtained segments. However, *RHRV* provides a function called *AnalyzeHRbyEpisodes* which performs this in one step: it makes use of an argument called *func* that will be a function to apply to the segments obtained. Results will be returned in two lists: *resultIn* and *resultOut*. Listing 6.4.2 shows and example where the standard deviation is estimated within *Apnea* and *Post-apnea* episodes.

R listing 6.4.2

```
SDApnea <-
   AnalyzeHRbyEpisodes(hrv.data, Tag = "Apnea",
   func = sd)
SDPostapnea <-
   AnalyzeHRbyEpisodes(hrv.data, Tag = "Post-apnea",
   func = sd)
cat("SD within apnea:", SDApnea$resultIn,
    "\nSD within post-apnea:", SDPostapnea$resultIn, "\n")

   ## SD within apnea: 9.835064
   ## SD within post-apnea: 2.279717
```

AnalyzePowerBandsByEpisodes is equivalent to *AnalyzeHRbyEpisodes*, but applying the function specified by *func* to the bands ULF, VLF, LF, and HF instead to the heart rate signal. *AnalyzePowerBandsByEpisodes* returns *resultIn* and *resultOut*, which will consist in two lists with parameters calculated by applying the specified function to the ULF, VLF, LF, and HF bands.

6.5 An Example

In this section, an example is included to show briefly the potential of episodes to study different temporal segments of a heart rate record. We will use the *HRVData* dataset included in the *RHRV* package. This record comes from a patient suffering from paraplegia and hypertension (systolic blood pressure above 200 mmHg). While obtaining the record, the patient is supplied with prostaglandin E1 (a vasodilator that is rarely employed), and systolic blood pressure fell to 100 mmHg for over an hour. Then, the blood pressure was slowly recovering until 150 mmHg, more or less.

This is by no means an exhaustive study, and it is only meant to illustrate the use of our software package. In Listing 6.5.1, this record is obtained, filtered, and interpolated. Episodes are added manually to reflect the different sections of the record, as can be seen in Fig. 6.5.

6.5 An Example

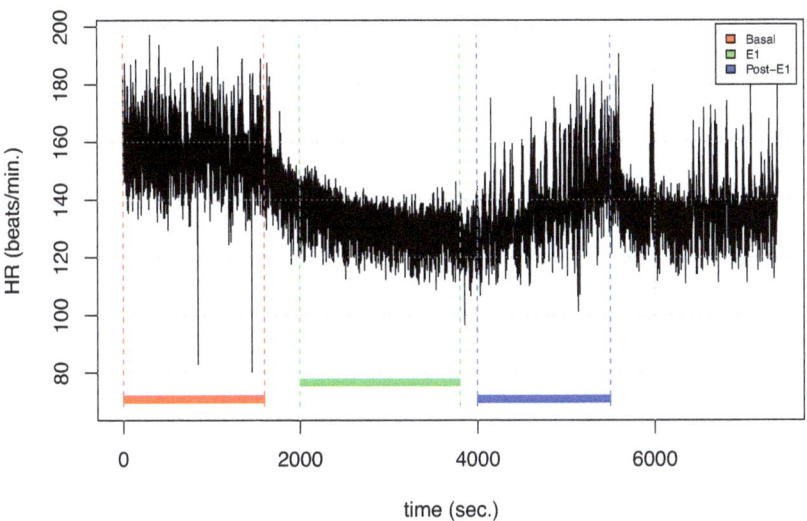

Fig. 6.5 Data included in RHRV with episodes added manually

R listing 6.5.1

```
library(RHRV)
data(HRVData)
hrv.data <- HRVData
hrv.data <- BuildNIHR(hrv.data)
hrv.data <- FilterNIHR(hrv.data)
hrv.data <- InterpolateNIHR(hrv.data)
hrv.data <- AddEpisodes(hrv.data,
   InitTimes = c(0, 2000, 4000),
   Durations = c(1600, 1800, 1500),
   Tags = c("Basal", "E1", "Post-E1"),
   Values = c(0, 0, 0))
PlotHR(hrv.data,Tags = "all")
```

Listing 6.5.2 presents a possible use of *AnalyzeHRbyEpisodes*. First, a function *CalPower* is defined to estimate the mean power of a segment, and then this function is passed as an argument to *AnalyzeHRbyEpisodes* to be applied to the segments resulting from dividing the heart rate signal according to the *Tag* argument.

R listing 6.5.2

```
CalPower <- function(v) {
   sumpower <- sum(v ^ 2) / length(v)
   return(sumpower)
}
rBasal <- AnalyzeHRbyEpisodes(hrv.data, Tag = "Basal",
   func= "CalPower")
rE1 <- AnalyzeHRbyEpisodes(hrv.data, Tag = "E1",
   func="CalPower")
cat("Mean power:\n  Basal situation:", rBasal$resultIn,
   "\n  After E1 prostaglandin:", rE1$resultIn,"\n")

## Mean power:
##   Basal situation: 25134.9
##   After E1 prostaglandin: 17590.71
```

Let us suppose that we want to compare visually the statistical properties of the LF/HF band in basal situation versus after administering prostaglandin E1. For this purpose, we can make use of the *SplitPowerBandByEpisodes* function to extract the values needed, and these data can be used in a boxplot, as it can be seen in Listing 6.5.3 and Fig. 6.6.

R listing 6.5.3

```
hrv.data <- CreateFreqAnalysis(hrv.data)
hrv.data <- CalculatePowerBand(hrv.data, size = 120,
   shift = 10)
dBasal <- SplitPowerBandByEpisodes(hrv.data,
   Tag = "Basal")
dE1 <- SplitPowerBandByEpisodes(hrv.data,Tag = "E1")
Basal_LFHF <-
   dBasal$InEpisodes$LF / dBasal$InEpisodes$HF
E1_LFHF <- dE1$InEpisodes$LF / dE1$InEpisodes$HF
boxplot(Basal_LFHF, E1_LFHF, outline = FALSE,
   col = c("blue", "red"),
   names = c("Basal", "E1"),
   main = "Basal vs prostaglandin E1",
   ylab = "LF/HF ratio")
```

6.6 Clinical Applications of HRV Analysis by Episodes

The importance of studying physiological events can be assessed by reviewing the literature. Moreover, some well-known challenges have been organized to study a particular type of episodes, such as the challenge from PhysioNet and Computers in Cardiology 2009 [9], dedicated to the analysis of acute hypotensive episodes, which can lead to the patient to have irreversible organ damage and death, or the Fourth

6.6 Clinical Applications of HRV Analysis by Episodes

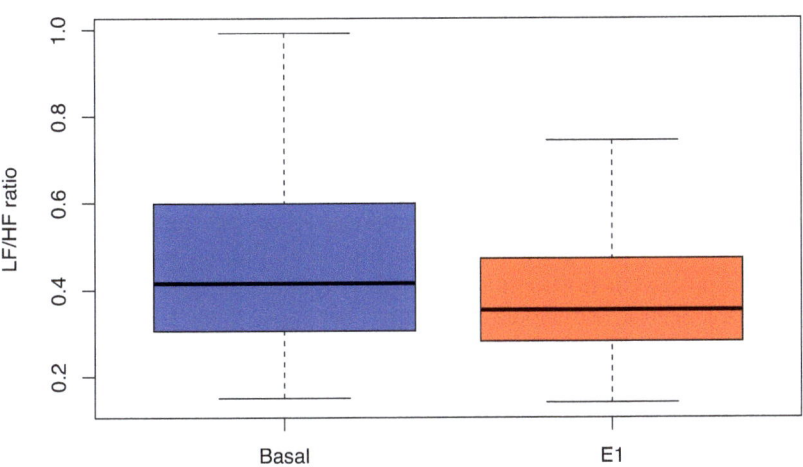

Fig. 6.6 LF/HF ratios for the basal and E1 episodes

China Physiological Signal Challenge 2021 [10], devoted to detect paroxysmal atrial fibrillation (PAFS) events, which is of great value for AF surgery, drug intervention, and the diagnosis and treatment of several clinical complications.

Many other scientific works include analysis of different types of episodes. Ling et al. [11] employed genetic-algorithms to detect hypoglycemic episodes (particularly dangerous at night) in diabetes mellitus patients. In a methodological review about the prevention of intradialytic hypotension, Sörnmo et al. [12] indicated that HRV analysis (linear and nonlinear parameters) in short successive heart rate segments during dialysis could provide useful information about these critical events.

Vila et al. [13] and Xing et al. [14] studied the spectral parameters of the heart ischemic episodes, and Mantaras et al. [15] analyzed the nonparametric and parametric time-frequency variables during the arousal from sleep. This physiological event is not abnormal, but produces a tachycardia-bradycardia pattern on the HRV parameters, which should be studied when arousal episodes become frequent and repetitive, since they can be an indicator of medical pathologies.

Romero-Legarreta et al. [16] analyzed the HRV in recurrent episodes of ventricular tachyarrhythmia, and Mohebbi at al. [17] concentrated on the prediction of paroxysmal atrial fibrillation employing both frequency and nonlinear analysis.

Apart from the specific events that may indicate physiological disorders in the patient, other interesting applications of the HRV analysis in different types of episodes can be found in literature. Several works indicate the influence of emotions in the HRV patterns. This is explained, for example, in an interesting paper [18] that studies the spectral parameters of the heart rate when the sample subjects are viewing diverse political spots, corresponding to different ideologies. Each spot can be considered and labeled as a different episode, and the corresponding heartbeat series segment can be analyzed, to obtain the influence of the distinct ideology spots on the subjects.

Studies about the influence of different types of emotional images in patients with mental disorder have also been performed. Hempel et al. [19] analyzed various types of images (that can be considered as episodes), corresponding to three categories (positive, erotic content; negative, physical injuries; and neutral, landscapes) in patients suffering from schizophrenia and control subjects, and concluded that the patients significantly increased the HR response when viewing positive pictures, compared to the control group. Rodríguez-Ruiz et al. [20] measured HRV parameters in bulimic women while they were viewing different types of images: food and pleasant, neutral, and unpleasant pictures, belonging to the IAPS (International Affective Picture System) database [21]. Poor autonomic regulation (low HRV) was found to play a modulatory influence in bulimic women.

Other studies about the HRV analysis have been devoted to the analysis of the relationship between music features and physiological patterns, such as the work by Orini et al. [22], who exposed healthy subjects to three types of episodes: pleasant music, Shepard tones (superposed sine waves separated by octaves), and unpleasant sounds. They found that the power of the HF modulation was lower when listening to pleasant music, compared to the unpleasant auditions.

References

1. A.L. Goldberger, L.A. Amaral, L. Glass, J.M. Hausdorff, P.C. Ivanov, R.G. Mark, J.E. Mietus, G.B. Moody, C.-K. Peng, H.E. Stanley, Physiobank, PhysioToolkit, and PhysioNet. Components of a new research resource for complex physiologic signals. Circulation **101**(23), e215–e220 (2000). [Online]. Available: https://doi.org/10.1161/01.cir.101.23.e215
2. G.B. Moody, R.G. Mark, A new method for detecting atrial fibrillation using RR intervals. Comput. Cardiol. **10**, 227–230 (1983)
3. Y. Ichimaru, G. Moody, Development of the polysomnographic database on CD-ROM. Psychiatry Clin. Neurosci. **53**(2), 175–177 (1999). [Online]. Available: https://doi.org/10.1046/j.1440-1819.1999.00527.x
4. L. Rodríguez-Liñares, P. Cuesta, R. Alonso, A. Méndez, M. Lado, X. Vila, VARVI: a software tool for analyzing the variability of the heart rate in response to visual stimuli, in *Computing in Cardiology Conference (CinC), 2013* (IEEE, Piscataway, 2013), pp. 401–404
5. The MathWorks Inc., Matlab version: 9.13.0 (r2022b), Natick, MA (2022). [Online]. Available: https://www.mathworks.com
6. C.A. García, A. Otero, X.A. Vila, M.J. Lado, L. Rodríguez-Liñares, J.M. Presedo, A.J. Méndez, Code for the book 'Heart Rate Variability Analysis with the R package RHRV' (Springer's use R!), https://github.com/RHRV-team/RHRVBook (2024)
7. PhysioBank. Last accessed 17 May 2024. [Online]. Available: http://www.physionet.org/physiobank
8. T. Penzel, G. Moody, R. Mark, A. Goldberger, J. Peter, The Apnea-ECG database, in *Computers in Cardiology 2000* (IEEE, Piscataway, 2000), pp. 255–258. [Online]. Available: https://doi.org/10.1109/CIC.2000.898505
9. Computers in Cardiology Challenge. Last accessed 17 May 2024. [Online]. Available: http://www.physionet.org/challenge/2009

References

10. Paroxysmal Atrial Fibrillation Events Detection from Dynamic ECG Recordings: The 4th China Physiological Signal Challenge 2021. Last accessed 26 Feb 2024. [Online]. Available: https://physionet.org/content/cpsc2021
11. S.S. Ling, H.T. Nguyen, Genetic-algorithm-based multiple regression with fuzzy inference system for detection of nocturnal hypoglycemic episodes. IEEE Trans. Inf. Technol. Biomed. **15**(2), 308–315 (2011). [Online]. Available: https://doi.org/10.1109/TITB.2010.2103953
12. L. Sörnmo, F. Sandberg, E. Gil, K. Solem, Noninvasive techniques for prevention of intradialytic hypotension. IEEE Rev. Biomed. Eng. **5**, 45–59 (2012). [Online]. Available: https://doi.org/10.1109/RBME.2012.2210036
13. J. Vila, F. Palacios, J. Presedo, M. Fernández-Delgado, P. Félix, S. Barro, Heart rate variability patterns in ischemic episodes, in *Computers in Cardiology 1997* (IEEE, Piscataway, 1997), pp. 371–374. [Online]. Available: https://doi.org/10.1109/CIC.1997.647910
14. W. Xing, X. Liang, S. Zhongwei, Y. Zibin, P. Yi, Heart rate variability analysis of ischemic and heart rate related ST-segment deviation episodes based on time-frequency method, in *International Conference on Functional Biomedical Imaging, 2007. NFSI-ICFBI 2007* (IEEE, Piscataway, 2007), pp. 162–164. [Online]. Available: https://doi.org/10.1109/NFSI-ICFBI.2007.4387715
15. M. Mantaras, M. Mendez, O. Villiantieri, N. Montano, V. Patruno, A. Bianchi, S. Cerutti, Non-parametric and parametric time-frequency analysis of heart rate variability during arousals from sleep, in *Computers in Cardiology, 2006* (IEEE, Piscataway, 2006), pp. 745–748
16. I.R. Legarreta, M. Reed, P. Addison, N. Grubb, G. Clegg, C. Robertson, J. Watson, Measurement of heart rate variability during recurrent episodes of ventricular tachyarrhythmia in one patient using wavelet transform analysis, in *Computers in Cardiology, 2004* (IEEE, Piscataway, 2004), pp. 469–472. [Online]. Available: https://doi.org/10.1109/CIC.2004.1442976
17. M. Mohebbi, H. Ghassemian, Prediction of paroxysmal atrial fibrillation based on non-linear analysis and spectrum and bispectrum features of the heart rate variability signal. Comput. Methods Programs Biomed. **105**(1), 40–49 (2012). [Online]. Available: https://doi.org/10.1016/j.cmpb.2010.07.011
18. Z. Wang, A.C. Morey, J. Srivastava, Motivated selective attention during political ad processing: the dynamic interplay between emotional ad content and candidate evaluation. Commun. Res. **41**, 119–156 (2014). [Online]. Available: https://doi.org/10.1177/00936502124417
19. R.J. Hempel, J.H. Tulen, N.J. van Beveren, H.G. van Steenis, P.G. Mulder, M.W. Hengeveld, Physiological responsivity to emotional pictures in schizophrenia. J. Psychiatric Res. **39**(5), 509–518 (2005). [Online]. Available: https://doi.org/10.1016/j.jpsychires.2004.11.004
20. S. Rodríguez-Ruiz, P.M. Guerra, S. Moreno, M.C. Fernández, J. Vila, Heart rate variability modulates eye-blink startle in women with bulimic symptoms. J. Psychophysiol. **26**(1), 10–19 (2012). [Online]. Available: https://doi.org/10.1027/0269-8803/a000064
21. P.J. Lang, M.M. Bradley, B.N. Cuthbert, International affective picture system (IAPS): affective ratings of pictures and instruction manual. Technical Report A-8 (2008)
22. M. Orini, R. Bailón, R. Enk, S. Koelsch, L. Mainardi, P. Laguna, A method for continuously assessing the autonomic response to music-induced emotions through HRV analysis. Med. Biol. Eng. Comput. **48**(5), 423–433 (2010). [Online]. Available: https://doi.org/10.1007/s11517-010-0592-3

Chapter 7
Putting It All Together: A Practical Example

Abstract This chapter presents an example of a complete HRV analysis of several long-duration ECG records employing the RHRV software package, including frequency, nonlinear, and time analysis. First, 3-h intervals of each ECG were extracted, corresponding to morning, afternoon, and night periods. Next, HRV analyses in both time and frequency domains were performed over each portion, and nonlinear values were also estimated. Finally, statistical analysis was applied over the variability parameters corresponding to these three types of fragments, to verify if differences existed among the morning, afternoon, and night ECG intervals. Some statistically significant differences were found between the morning and night periods. In particular, *HRVi* (*HRV index*, time) and *Poincaré* SD_2 (nonlinear) parameters differ in a statistical way.

7.1 Problem Statement

Circadian rhythm has been widely studied since several decades ago. It can be defined as variations in the physiological functions that occur in a period of 1 day [1]. Light is the most important external factor when synchronizing the circadian rhythm, and it can be altered by several factors, such as work shifts, journeys, or even season cycles [2].

Much research about the circadian rhythm and its influence in humans can be found in the literature. Several studies analyzed the influence of the amount of daylight and darkness in different seasons and at different places on the planet, over the circadian clocks and corresponding periods of sleep and wakefulness [3]. Other investigators have devoted their efforts to study the circadian rhythm in blind people [4], during sleep periods [5], or in patients suffering from asthma [6].

It also well-known that morning hours are associated with a high risk of suffering from cardiovascular events [7]. In fact, several works can be found in the literature which show that healthy subjects typically begin the night with high values of HRV parameters that decrease through the night until morning, and these low values can be interpreted as a sign of cardiovascular risk, concluding that cardiovascular events are more likely during the last hours of night [8, 9].

There are many studies that try to establish the relationship between the circadian rhythm and HRV. A significant difference between day and night HRV was found by Ramaekers et al. [10], reflecting a higher vagal modulation during the night. Massin et al., while studying the circadian variations in frequency and time domains in children, found a rise during sleep, with the exception of the LF/HF ratio, which increased during daytime [11]. Malik et al. also related circadian rhythm and HRV and analyzed the influence of the circadian rhythm after acute myocardial infarction on the prognosis value of HRV [12]. The circadian rhythm of the HRV in patients suffering from coronary artery disease was studied by Huikuri et al. [13], and results were compared to those obtained by healthy subjects. No significant circadian rhythms in spectral components of HRV were observed in patients with coronary artery disease; the night-day difference in LF/HF ratio was smaller in those patients than in the healthy subjects.

Vanoli et al. measured HRV from 5 min of continuous ECG recording in patients with and without coronary artery disease and found a LF/HF ratio that decreased from the awake state to non-REM sleep in healthy subjects [14].

As it was stated before, Boudreau et al. [7] investigated the interaction between circadian and sleep-wake-dependent processes on HRV, analyzing the different frequency components of HRV, finding a significant interaction. The darkness rise was associated with a rapid increase in the average RR interval and cardiac parasympathetic modulation, and a reduction in the LF band, as well as in the LF/HF ratio, was also found.

Other authors have studied the influence of factors such as sex, gender, and night variations over the nonlinear parameters of HRV [15]. They analyzed the fractal dimension and DFA of HRV, as well as the correlation dimension, the approximate entropy, and the Lyapunov exponent. A higher nonlinear behavior was observed during the night versus day, and nonlinear heart rate fluctuations decreased with age.

In this chapter, we will analyze, employing the RHRV software package, the HRV indices in time and frequency domains, as well as nonlinear parameters, in three different periods of the circadian rhythm corresponding to morning, afternoon, and night. Even although we have used real ECG recordings, readers should have in mind that the main purpose of this work is to give a detailed explanation about the options and uses of the software tool, and we do not pretend to have designed and applied a rigorous clinical procedure to test an initial hypothesis. Then, results obtained in this example should be cautiously considered and interpreted.

7.2 Methodology

7.2.1 Database Description

In order to establish if there are significant differences in the HRV parameters across different time periods of the circadian rhythm corresponding to 3-h intervals in the morning, afternoon, and night, ten ECG records were selected from the Long-Term ST Database [16, 17], a public database composed of 86 24-h ECG recordings of 80 human subjects that can be freely downloaded from the PhysioBank website [18]. Each recording is between 21 and 24 h in duration and may contain more than one ECG signal, even though only the first signal in each case was considered in this study.

From the set of files associated with each record, we are interested in the following ones: a .hea file, which is a text file including header information (ECG files associated with the patient, clinical data, ECG channels, sample frequency, calibration constants, record length, diagnoses, medical history, treatment, time, and date information), and a .atr binary file, which contains manually annotated beat positions.

Listing 7.2.1 shows the R commands that may be used to download the files associated with the ECG recordings selected for this study.

R listing 7.2.1

```
# HRVData structure containing the heart beats
OriginDir <-
  "http://www.physionet.org/physiobank/database/ltstdb/"
items = c("s20011", "s20021", "s20031", "s20041",
          "s20071", "s20091", "s20111", "s20121",
          "s20131", "s20141")
for (i in 1:length(items)) {
  headerFile <- paste(items[[i]], ".hea", sep = "")
  annotationFile <- paste(items[[i]], ".atr", sep = "")

  files <- c(headerFile, annotationFile)
  OriginFiles <- paste(OriginDir, files, sep = "")
  for (j in 1:2)
    download.file(OriginFiles[j], files[j])
}
```

Table 7.1 shows the recordings employed in this example, the starting time (in hh:mm:ss), and the initial time intervals (in seconds) selected in each case. As said before, the interval duration is 3 h (10,800 s) in all cases. It must be noticed that the first labeled episode corresponds to the afternoon, and not to the morning, as expected. This is due to the fact that all the ECG recordings started between 9:00 and 17:08 h, and thus, the first ECG recording portion corresponded to the afternoon hours. In all cases, morning intervals were obtained by selecting a 3-h period between 6:00 a.m. and 10:00 a.m. The afternoon period started, depending

Table 7.1 Selection of patients and intervals corresponding to morning, afternoon, and night periods. Starting time is given in hh:mm:ss and the intervals in seconds

Patient	Starting recording	Initial afternoon	Initial night	Initial morning
s20011	17:08:00	1320	24,720	49,920
s20021	11:00:00	14,400	43,200	68,400
s20031	09:24:00	23,760	52,560	74,160
s20041	12:06:00	14,040	42,480	67,680
s20071	12:51:00	11,340	40,140	65,340
s20091	13:01:00	10,740	39,540	64,740
s20111	09:10:00	24,600	53,400	75,000
s20121	09:21:00	23,940	52,740	74,340
s20131	13:43:00	8220	37,020	62,220
s20141	13:16:00	13,440	42,240	67,440

on the initial ECG time recording, between 16:00 and 17:30 h and never finished after 20:30 h. Finally, the 3-h night intervals were selected from 23:00 to 24:00 h and finished between 02:00 and 03:00 a.m.

7.2.2 Applying HRV Analysis

The steps to perform both frequency and time analysis, as well as the study of nonlinear parameters, are the following:

1. Creation of the *HRVData* structure.
2. Original signal loading.
3. Preprocessing analysis: including instantaneous heart rate signal extraction, artifact rejection, estimation of RR intervals, filtering, and interpolation of the heart rate, as explained in Chap. 2.
4. Extraction of the 3-h intervals: signal intervals corresponding to morning, afternoon, and night periods were extracted for each patient.
5. Frequency analysis: spectral parameters were obtained for the three intervals of the signal. Specifically, values for the HF and LF spectral bands, as well as the HRV total power, were calculated. The LF/HF ratio was also included in the spectral analysis for each type of interval (morning, afternoon, and night).
6. Time analysis: values for some of the time variables were calculated for each specified time interval. In this way, we have analyzed the following variables: *pNN50* (percentage of differences between adjacent intervals that are greater than 50 ms), *r-MSSD* (square root of the mean of the squares of the differences between adjacent intervals), and *HRVi* (integral of the density distribution divided by the maximum of the density distribution).
7. Nonlinear analysis: some nonlinear parameters were also estimated for each interval of the original signal. In particular, sample entropy (used for assessing

7.2 Methodology

the complexity of a physiological time-series signal), and the Poincaré plot (a nonlinear measure used for quantifying the self-similarity of signals, in which each RR interval is plotted versus the next RR interval) were calculated.

Listing 7.2.2 presents the R code employed to perform the corresponding analyses to the signal belonging to patient "s20011." For the rest of cases, the specific morning, afternoon, and night intervals described in Table 7.1 were used.

R listing 7.2.2

```
# hrv.data structure creation and preprocessing

# Index for the record to be analyzed:
# from 1 (s20011) to 10 (s20141)
i = 1

hrv.data <- CreateHRVData()
hrv.data <- SetVerbose(hrv.data, TRUE)

hrv.data <- LoadBeat("WFDB", HRVData, "s20011",
                    annotator = "atr")

hrv.data <- BuildNIHR(hrv.data)
hrv.data <- FilterNIHR(hrv.data)
hrv.data <- InterpolateNIHR(hrv.data)

# Intervals definition and HRVData structures created

duration <- 10800
initAfternoon <- 1320
initNight <- 24720
initMorning <- 49920

hrv.data.morning <-
  ExtractTimeSegment(hrv.data,
                     starttime = initMorning,
                     endtime = initMorning + duration)
hrv.data.afternoon <-
  ExtractTimeSegment(hrv.data,
                     starttime = initAfternoon,
                     endtime = initAfternoon + duration)
hrv.data.night <-
  ExtractTimeSegment(hrv.data,
                     starttime = initNight,
                     endtime = initNight + duration)

# Spectral analysis

hrv.data.morning <-
  CreateFreqAnalysis(hrv.data.morning)
hrv.data.morning <-
  CalculatePowerBand(hrv.data.morning,
                     indexFreqAnalysis = 1,
```

```
                            shift = 30)

hrv.data.afternoon <-
  CreateFreqAnalysis(hrv.data.afternoon)
hrv.data.afternoon <-
  CalculatePowerBand(hrv.data.afternoon,
                      indexFreqAnalysis = 1,
                      shift = 30)

hrv.data.night <-
  CreateFreqAnalysis(hrv.data.night)
hrv.data.night <-
  CalculatePowerBand(hrv.data.night,
                      indexFreqAnalysis = 1,
                      shift = 30)

# Time analysis

hrv.data.morning <- CreateTimeAnalysis(hrv.data.morning,
                                        size = 300)
hrv.data.afternoon <- CreateTimeAnalysis(hrv.data.afternoon,
                                          size = 300)
hrv.data.night <- CreateTimeAnalysis(hrv.data.night,
                                      size = 300)

# Nonlinear analysis

hrv.data.morning <-
  CreateNonLinearAnalysis(hrv.data.morning)
hrv.data.morning <-
  CalculateCorrDim(hrv.data.morning,
                    indexNonLinearAnalysis = 1,
                    minEmbeddingDim = 2,
                    maxEmbeddingDim = 8, timeLag = 1,
                    minRadius = 1, maxRadius = 15,
                    pointsRadius = 20, theilerWindow = 10,
                    corrOrder = 2, doPlot = FALSE)
hrv.data.morning <- CalculateSampleEntropy(hrv.data.morning)
hrv.data.morning <- EstimateSampleEntropy(hrv.data.morning)
hrv.data.morning <- PoincarePlot(hrv.data.morning)

hrv.data.afternoon <-
  CreateNonLinearAnalysis(hrv.data.afternoon)
hrv.data.afternoon <-
  CalculateCorrDim(hrv.data.afternoon,
                    indexNonLinearAnalysis = 1,
                    minEmbeddingDim = 2,
                    maxEmbeddingDim = 8, timeLag =1 ,
                    minRadius = 1, maxRadius = 15,
                    pointsRadius = 20, theilerWindow = 10,
                    corrOrder = 2, doPlot = FALSE)
hrv.data.afternoon <- CalculateSampleEntropy(hrv.data.afternoon)
hrv.data.afternoon <- EstimateSampleEntropy(hrv.data.afternoon)
```

7.2 Methodology

```
hrv.data.afternoon <- PoincarePlot(hrv.data.afternoon)

hrv.data.night <-
  CreateNonLinearAnalysis(hrv.data.night)
hrv.data.night <-
  CalculateCorrDim(hrv.data.night,
                  indexNonLinearAnalysis = 1,
                  minEmbeddingDim=2,
                  maxEmbeddingDim = 8, timeLag = 1,
                  minRadius = 1, maxRadius = 15,
                  pointsRadius = 20, theilerWindow =10,
                  corrOrder = 2, doPlot = FALSE)
hrv.data.night <- CalculateSampleEntropy(hrv.data.night)
hrv.data.night <- EstimateSampleEntropy(hrv.data.night)
hrv.data.night <- PoincarePlot(hrv.data.night)


# new morning, afternoon and data strcutures creation

MorningData[[i]] <- hrv.data.morning
AfternoonData[[i]] <- hrv.data.afternoon
NightData[[i]] <- hrv.data.night
```

This code is explained in the following paragraphs. The first lines of Listing 7.2.2, labeled as *hrv.data structure creation and preprocessing*, create the *hrv.data* structure and then preprocessing is applied. As a result, the interpolated heart rate signal is obtained.

After that, a different *HRVData* structure is created for each of the three 3-h periods to be analyzed, corresponding to morning, afternoon, and night intervals (*hrv.data.morning, hrv.data.afternoon, hrv.data.night*), employing the function *ExtractTimeSegment* (see Listing 7.2.2, under the *Intervals definition and HRVData structures created* comment).

The next step is the calculation of spectral HRV parameters for each interval of the signal (code lines labeled *Spectral analysis*). A frequency analysis was created for each *HRVData* structure (morning, afternoon, and night), employing the *CreateFreqAnalysis* function. The corresponding spectral power bands were obtained executing the function *CalculatePowerBand*, employing the STFT over a zero mean segment of the signal, using window size and shift values of 300 and 30 s, respectively.

After the frequency analysis, the time analysis was performed for each type of interval; this was labeled in the code as *Time analysis*. The *CreateTimeAnalysis* function was applied, selecting a window size of 300 s.

Finally, nonlinear parameters were obtained in the portion of the code labeled *Nonlinear analysis*. In this case, a nonlinear analysis has to be created previously using the function *CreateNonLinearAnalysis*.

After that, the sample entropy was calculated with the functions *CalculateSampleEntropy* and *EstimateSampleEntropy*. Poincaré analysis was also applied through the function *PoincarePlot*, and the ellipse fitting technique was applied. The SD_1 (minor axis) and SD_2 (major axis) ellipse parameters were calculated.

The code presented in Listing 7.2.2 was executed for each of the recordings. As a result, three different lists (one for each time period, i.e., morning, afternoon, and night, and labeled as *MorningData*, *AfternoonData*, and *NightData*), were obtained. Each list contained ten elements (belonging to the ten selected cases), and each element corresponded to a *HRVData* structure with all the previous calculations: heart rate values, beats, frequency, time, and nonlinear analysis. For example, *MorningData[[1]]* contained all the information related to the 3-h morning period for recording "s20011."

The three sets of values for all ECG recordings (*MorningData*, *AfternoonData*, and *NightData*) were stored in three different lists for morning, afternoon, and night periods (*Morning*, *Afternoon*, and *Night*, respectively). As an example, Listing 7.2.3 provides the R code for the generation of the *Morning* list.

R listing 7.2.3

```
Morning <- list()
# One index for each of the values of each analyzed record
for (i in 1:10) {
  Morning$HRV[[i]] <-
    mean(MorningData[[i]]$FreqAnalysis[[1]]$HRV)
  Morning$LF[[i]] <-
    mean(MorningData[[i]]$FreqAnalysis[[1]]$LF)
  Morning$HF[[i]] <-
    mean(MorningData[[i]]$FreqAnalysis[[1]]$HF)
  Morning$LFHF[[i]] <-
    mean(MorningData[[i]]$FreqAnalysis[[1]]$LFHF)

  Morning$pNN50[[i]] <-
    MorningData[[i]]$TimeAnalysis[[1]]$pNN50
  Morning$rMSSD[[i]] <-
    MorningData[[i]]$TimeAnalysis[[1]]$rMSSD
  Morning$HRVi[[i]] <-
    MorningData[[i]]$TimeAnalysis[[1]]$HRVi

  Morning$SampleEntropy[[i]] <-
    MorningData[[i]]$NonLinearAnalysis[[1]]$sampleEntropy$statistic
  Morning$PoincareSD1[[i]] <-
    MorningData[[i]]$NonLinearAnalysis[[1]]$PoincarePlot$SD1
  Morning$PoincareSD2[[i]] <-
    MorningData[[i]]$NonLinearAnalysis[[1]]$PoincarePlot$SD2
}
```

From the lists *Morning*, *Afternoon*, and *Night*, average values were calculated for all the parameters in both frequency and time domain, as well as nonlinear parameters, for the three types of 3-h intervals, for all the patients participating in the study. Results are presented in Table 7.2.

The main goal of this study was to determine if statistical significant differences existed among HRV values for the three types of episodes. To this end, a statistical

7.2 Methodology

Table 7.2 Average values and corresponding standard deviations for the spectral, time, and nonlinear parameters corresponding to morning, afternoon, and night periods

	Morning	Afternoon	Night
HRV	1215.29 ± 594.05	800.21 ± 346.75	765.14 ± 569.52
HF	111.78 ± 167.31	68.95 ± 72.08	106.09 ± 200.77
LF	183.97 ± 144.02	130.06 ± 75.72	121.94 ± 92.66
LF/HF	4.18 ± 3.62	4.27 ± 3.33	2.93 ± 2.63
pNN50	4.18 ± 3.62	3.47 ± 4.39	6.85 ± 10.07
r-MSSD	29.81 ± 14.27	24.71 ± 8.82	29.45 ± 13.21
HRVi	23.40 ± 7.10	18.94 ± 6.38	14.72 ± 4.88
Sample entropy	1.38 ± 0.24	1.33 ± 0.23	1.44 ± 0.21
Poincaré SD1	21.08 ± 10.09	17.48 ± 6.23	20.83 ± 9.34
Poincaré SD2	139.67 ± 46.01	104.65 ± 31.37	93.33 ± 36.81

paired t-test was conducted. This statistical analysis was applied to all the parameters studied and between all the types of episodes, and the 95% confidence intervals (95% CIs) and *p*-values were also calculated.

Listing 7.2.4 presents, as an example, the R code for comparison between morning and afternoon periods for spectral, time, and nonlinear parameters.

It should be noted that Morning and Afternoon are R lists, which means that they contain an ordered collection of objects, each one corresponding to one of the statistics selected in the study, and contains specific fields for HRV, *HF*, *LF*, *LF/HF*, *pNN50*, *r-MSSD*, *HRVi*, *SampleEntropy*, and *Poincaré* values. Moreover, the t-test for a particular parameter x was stored in a variable called xMorningAfternoon, with x corresponding to the parameter under study.

R listing 7.2.4

```
HRVMorningAfternoon <- t.test(Morning$HRV,
                              Afternoon$HRV,
                              paired = TRUE)
HFMorningAfternoon <- t.test(Morning$HF,
                             Afternoon$HF,
                             paired = TRUE)
LFMorningAfternoon <- t.test(Morning$LF,
                             Afternoon$LF,
                             paired = TRUE)
LFHFMorningAfternoon <- t.test(Morning$LFHF,
                               Afternoon$LFHF,
                               paired = TRUE)
pNN50MorningAfternoon <- t.test(Morning$pNN50,
                                Afternoon$pNN50,
                                paired = TRUE)
rMSDMorningAfternoon <- t.test(Morning$rMSD,
                               Afternoon$rMSD,
                               paired = TRUE)
HRViMorningAfternoon <- t.test(Morning$HRVi,
                               Afternoon$HRVi,
```

```
                            paired = TRUE)
MorningAfternoon <- t.test(Morning$SampleEntropy,
                           Afternoon$SampleEntropy,
                           paired = TRUE)
PoincareSD1MorningAfternoon <-
  t.test(Morning$PoincareSD1, Afternoon$PoincareSD1,
         paired = TRUE)
PoincareSD2MorningAfternoon <-
  t.test(Morning$PoincareSD2, Afternoon$PoincareSD2,
         paired = TRUE)
```

After applying t-tests, 95% CI intervals and *p*-values were obtained in each case. A general statistical condition for two independent sets to be significantly different with respect to a given variable is that the difference in the variable values between both independent sets does not assume a zero value within the corresponding 95% confidence interval [19].

In our case, some statistically significant differences were found for several parameters and intervals. Table 7.3 presents all the values obtained for the *p*-values and corresponding 95% confidence intervals for the differences between all the types of episodes. Significant parameters (*p*-value obtained applying the Bonferroni correction) were marked with an asterisk * (*p*-value < 0.02).

From Table 7.3, it can be noticed that four statistically significant differences were found for the three types of episodes: (1) *HRV* total power when comparing morning versus both afternoon and night episodes and (2) *HRVi* and *Poincaré* SD_2 (major axis) parameters differ in a statistical way between morning and night intervals.

As it was stated in Sect. 7.1, we were not pretending to extract conclusions and clinical information about the differences in HRV values between different episodes;

Table 7.3 Significant analysis comparing the three types of periods

	Morning versus afternoon		Morning versus night		Afternoon versus night	
	p-value	95%CI	*p*-value	95%CI	*p*-value	95%CI
HRV	0.02*	(74.29, 755.83)*	0.003*	(191.79, 708.49)*	0.81	(−300.66, 370.80)
HF	0.30	(−46.43, 132.01)	0.84	(−58.56, 69.94)	0.49	(−152.95, 78.67)
LF	0.23	(−40.72, 148.54)	0.13	(−24.17, 148.22)	0.80	(−62.80, 79.04)
HF/LF	0.50	(−2.39, 1.27)	0.32	(−1.46, 3.99)	0.07	(−0.17, 3.83)
pNN50	0.03	(0.31, 5.34)	0.65	(−3.25, 2.14)	0.11	(−7.68, 0.91)
r-MSSD	0.19	(−3.17, 13.36)	0.91	(−6.38, 7.09)	0.18	(−12.20, 2.71)
HRVi	0.14	(−1.82, 10.75)	0.007*	(2.96, 14.39)*	0.07	(−0.48, 8.91)
Sample entropy	0.68	(−0.20, 0.29)	0.47	(−0.22, 0.11)	0.22	(−0.28, 0.07)
Poincaré SD_1	0.19	(−2.24, 9.45)	0.91	(−4.51, 5.01)	0.18	(−8.62, 1.91)
Poincaré SD_2	0.06	(−1.81, 71.87)	0.0004*	(27.02, 65.66)*	0.45	(−21.19, 43.82)

we were just focusing on explaining how the RHRV software package can be used to analyze the variability of heart rate signals. Because of this, we will not try to clinically understand our results and give a rigorous explanation. Thus, these results should be considered and interpreted with caution.

References

1. R.T.W.L. Conroy, J.N. Mills, *Human Circadian Rhythms* (Churchill, London, 1970)
2. D.S. Minors, J.M. Waterhouse, *Circadian Rhythms and the Human* (Butterworth-Heinemann, Oxford, 2013)
3. D. Purves, G.J. Augustine, D. Fitzpatrick, L.C. Katz, A.-S. Lamantia, J.O. McNamara, S.M. Williams, *Neuroscience* (Sinauer, Sunderland, 2001)
4. R.L. Sack, A.J. Lewy, M.L. Blood, L.D. Keith, H. Nakagawa, Circadian rhythm abnormalities in totally blind people: incidence and clinical significance. J. Clin. Endocrinol. Metab. **75**(1), 127–134 (1992) [Online]. Available: https://doi.org/10.1210/jcem.75.1.1619000
5. C.R. Jones, S.S. Campbell, S.E. Zone, F. Cooper, A. DeSano, P.J. Murphy, B. Jones, L. Czajkowski, L.J. Ptáček, Familial advanced sleep-phase syndrome: a short-period circadian rhythm variant in humans. Nat. Med. **5**(9), 1062–1065 (1999). [Online]. Available: https://doi.org/10.1038/12502
6. M. Hetzel, T. Clark, Comparison of normal and asthmatic circadian rhythms in peak expiratory flow rate. Thorax **35**(10), 732–738 (1980). [Online]. Available: https://doi.org/10.1136/thx.35.10.732
7. P. Boudreau, W.H. Yeh, G.A. Dumont, D.B. Boivin, A circadian rhythm in heart rate variability contributes to the increased cardiac sympathovagal response to awakening in the morning. Chronobiol. Int. **29**(6), 757–768 (2012). [Online]. Available: https://doi.org/10.3109/07420528.2012.674592
8. K. Kario, T.G. Pickering, Y. Umeda, S. Hoshide, Y. Hoshide, M. Morinari, M. Murata, T. Kuroda, J.E. Schwartz, K. Shimada, Morning surge in blood pressure as a predictor of silent and clinical cerebrovascular disease in elderly hypertensives a prospective study. Circulation **107**(10), 1401–1406 (2003). [Online]. Available: https://doi.org/10.1161/01.CIR.0000056521.67546.AA
9. W.B. White, Cardiovascular risk and therapeutic intervention for the early morning surge in blood pressure and heart rate. Blood Pressure Monit. **6**(2), 63–72 (2001). [Online]. Available: https://doi.org/10.1097/00126097-200104000-00001
10. D. Ramaekers, H. Ector, A. Aubert, A. Rubens, F. Van de Werf, Heart rate variability and heart rate in healthy volunteers. Eur. Heart J. **19**, 1334–41 (1998). [Online]. Available: https://doi.org/10.1053/euhj.1998.1084
11. M.M. Massin, K. Maeyns, N. Withofs, F. Ravet, P. Gérard, Circadian rhythm of heart rate and heart rate variability. Arch. Dis. Childhood **83**(2), 179–182 (2000). [Online]. Available: https://doi.org/10.1136/adc.83.2.179
12. M. Malik, T. Farrell, A.J. Camm, Circadian rhythm of heart rate variability after acute myocardial infarction and its influence on the prognostic value of heart rate variability. Am. J. Cardiol. **66**(15), 1049–1054 (1990). [Online]. Available: https://doi.org/10.1016/0002-9149(90)90503-S
13. H.V. Huikuri, M.K. Linnaluoto, T. Seppänen, K.J. Airaksinen, K.M. Kessler, J.T. Takkunen, R.J. Myerburg, Circadian rhythm of heart rate variability in survivors of cardiac arrest. Am. J. Cardiol. **70**(6), 610–615 (1992). [Online]. Available: https://doi.org/10.1016/0002-9149(92)90200-I
14. E. Vanoli, P.B. Adamson, Ba-Lin, G.D. Pinna, R. Lazzara, W.C. Orr, Heart rate variability during specific sleep stages. A comparison of healthy subjects with patients after myocardial

infarction. Circulation **91**(7), 1918–1922 (1995). [Online]. Available: https://doi.org/10.1161/01.CIR.91.7.1918
15. S. Vandeput, B. Verheyden, A. Aubert, S. Van Huffel, Nonlinear heart rate dynamics: circadian profile and influence of age and gender. Med. Eng. Phys. **34**(1), 108–117 (2012). [Online]. Available: https://doi.org/10.1016/j.medengphy.2011.07.004
16. A.L. Goldberger, L.A. Amaral, L. Glass, J.M. Hausdorff, P.C. Ivanov, R.G. Mark, J.E. Mietus, G.B. Moody, C.-K. Peng, H.E. Stanley, PhysioBank, PhysioToolkit, and PhysioNet. Components of a new research resource for complex physiologic signals. Circulation **101**(23), e215–e220 (2000). [Online]. Available: https://doi.org/10.1161/01.CIR.101.23.e215
17. F. Jager, A. Taddei, G.B. Moody, M. Emdin, G. Antolič, R. Dorn, A. Smrdel, C. Marchesi, R.G. Mark, Long-term ST database: a reference for the development and evaluation of automated ischaemia detectors and for the study of the dynamics of myocardial ischaemia. Med. Biol. Eng. Comput. **41**(2), 172–182 (2003). [Online]. Available: https://doi.org/10.1007/BF02344885
18. PhysioBank. Last accessed 26 Feb 2024 [Online]. Available: http://www.physionet.org/physiobank
19. A. Grafen, R. Hails, *Modern Statistics for the Life Sciences* (Oxford University Press, Oxford, 2002)

Chapter 8
Automating HRV Analysis: *RHRVEasy*

Abstract The previous chapter presented a typical HRV analysis. It involved calculating a series of HRV indices for various time intervals. This analysis relied on loops to process each interval, which repeated the same calculations numerous times. Subsequently, statistical tests were employed to identify differences between the indices calculated for the different time intervals. This repetitive approach is a widespread pattern in HRV analysis. With the development of a tool such as RHRV, the next logical step was to automate these repetitive tasks, streamlining the entire HRV analysis process, including both the calculation of the HRV indices and the statistical analysis. This is the goal of *RHRVEasy*.

8.1 Introduction

The final objective of HRV studies often involves identifying statistically significant differences in HRV indices between two and more populations. These indices may serve as potential markers of a pathology or support the interpretation of the underlying physiological processes in different populations. Consequently, HRV analyses typically involve calculating indices for each recording within each population, followed by statistical analysis to detect interpopulation differences. The choice of statistical test depends on the normality of each index and the number of groups being compared. Furthermore, since multiple indices are often calculated (up to 31 in the case of RHRV), it is necessary to correct the significance level of the statistical tests to account for the multiple comparisons. Finally, in studies involving more than two experimental groups where significant differences were found, a post hoc analysis is also needed to find out which groups are different.

With a tool such as the RHRV package, based on programming and not on a graphical user interface, it is possible to automate the calculation of the indices for all recordings from all populations. Developing the necessary code to automate the typical statistical analysis described in the previous paragraph would also be possible. This is the purpose of *RHRVEasy*. This functionality is the main novelty of version 5.0 of RHRV and of the second edition of the present book.

8.2 Time and Frequency Index Calculation with *RHRVEasy*

The main *RHRVEasy* function requires a single mandatory parameter: a list with the folders that contain the recordings of each experimental group (see Listing 8.2.1). This list must include at least two folders. Each folder must contain the RR recordings for a single experimental group exclusively. No other files should be present within these folders, as *RHRVEasy* will attempt to load all files it finds. The names of each folder will be used within *RHRVEasy* to identify the corresponding experimental group.

An optional *format* parameter specifies the format of the RR interval files. All formats supported by the RHRV package are available, including WFDB, ASCII, EDFPlus, RR, Polar, Suunto, and Ambit. By default, the format is assumed to be RR, where beat distances in seconds are stored in a single column of an ASCII file. A verbose mode that provides detailed information about the different steps can be activated by setting *verbose* to *TRUE* (see Listing 8.2.1).

R listing 8.2.1

```
# Prototype of the main RHRVEasy function.
RHRVEasy <-
  function(folders,
           verbose = FALSE,
           format = "RR",
           typeAnalysis = "fourier",
           correctionMethod = "bonferroni",
           significance = 0.05,
           nonLinear = FALSE,
           doRQA = FALSE,
           nJobs = 1,
           saveHRVIndicesInPath = NULL,
           ...)
```

As we have seen throughout this book, any HRV analysis involves many configuration parameters. Precisely, one of the advantages of RHRV is that these parameters can be documented in a code that can be easily shared, supporting the simple replication of the results of an analysis. While *RHRVEasy* uses sensible default parameters (the main ones will be documented in the following sections), it is possible to override any of them through the R special ... argument. If any of these parameters is specified, its value will be used in the calculations. If not, a default value will be used.

The *RHRVEasy* function follows a well-defined workflow, as shown in Fig. 8.1, which we detail in the following paragraphs.

Upon being called, the *RHRVEasy* function starts by initializing the necessary data structures required to perform HRV analysis, as shown in Fig. 8.2. First, it validates the *folders* parameter, ensuring that the specified folders exist and contain files with RR recordings. If any problems arise, an error message is displayed.

8.2 Time and Frequency Index Calculation with *RHRVEasy*

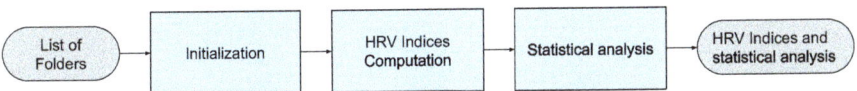

Fig. 8.1 Overview of the operations performed by the *RHRVEasy* package. Rectangles with rounded edges represent *data.frames* or *lists*, whereas rectangles with squared edges represent operations, which may be composed of several steps

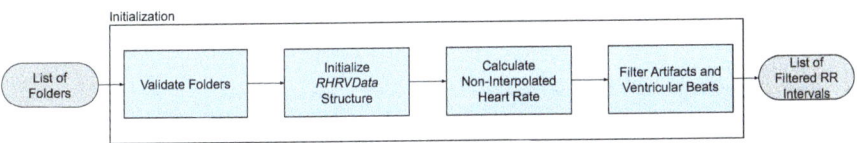

Fig. 8.2 Workflow with the operations followed during the initialization step of *RHRVEasy*

Next, it iterates through each recording within each experimental group using a loop. During each iteration, a data structure to load the beats of each recording is created using the *CreateHRVData()* function. Then the RR intervals are loaded (*LoadBeat()*), the non-interpolated heart rate is calculated (*BuildNIHR()*), and artifacts and ventricular beats are filtered (*FilterNIHR()*). All these functions are called with the RHRV default parameters unless they have been overridden in the *RHRVEasy* function call.

During this initialization a *data.frame* named *HRVIndices* is also created. It will be used to store in each row the HRV indices of each recording (see Fig. 8.3). This *data.frame* also includes two additional columns. One column contains the file name from which the RR intervals were loaded. The other specifies the experimental group to which the data belongs (which is the folder name where the file was located). With the RR intervals loaded and the necessary data structures initialized, the calculation of HRV indices can now start. The workflow for computing the HRV indices is detailed in Fig. 8.4.

The calculation of the HRV indices begins with the time-domain indices using the *CreateTimeAnalysis()* function. For each recording, it computes SDNN, SDANN, SDNNIDX, pNN50, SDSD, rMSSD, IRRR, MADRR, TINN, and HRVi (see Sect. 3.1). The default window size for time analysis in *RHRVEasy* is 300 s (*size* parameter), and the histogram bin width (*interval* parameter) for the TINN index calculation is 7.8125 ms.

In short RR recordings, it may not be possible to calculate SDANN and SDNNIDX. SDANN is the standard deviation of mean RR intervals within each window (300 s by default), while SDNNIDX is the average of RR standard deviations across windows (see Sect. 3.1). For recordings with only one window, or when there are windows only containing missing data, it is not possible to calculate these indices. If SDANN and SDNNIDX cannot be computed, *RHRVEasy* sets their values to NA (missing data), and if the verbose mode is activated, it generates a warning message to alert the user.

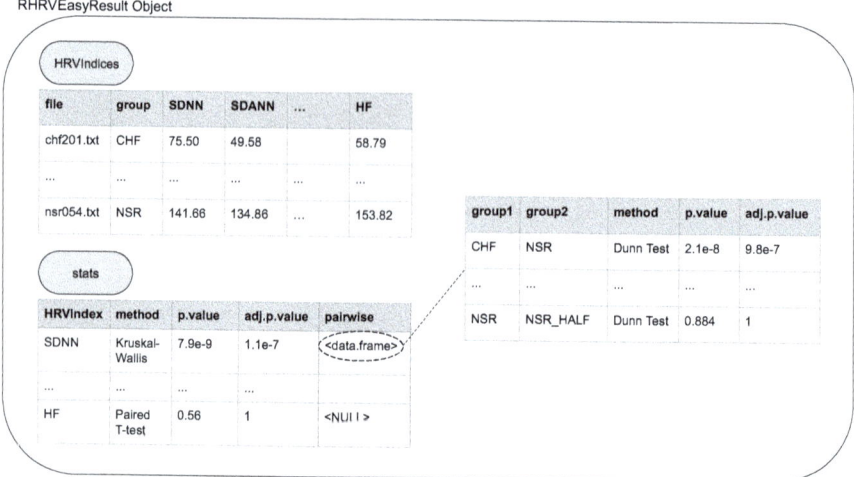

Fig. 8.3 Representation of the *RHRVEasyResult* object used by *RHRVEasy* to store both the computed HRV indices and the results from the statistical analysis. The object consists of two slots named *HRVIndices* and *stats*, as noted by the rectangles with rounded edges

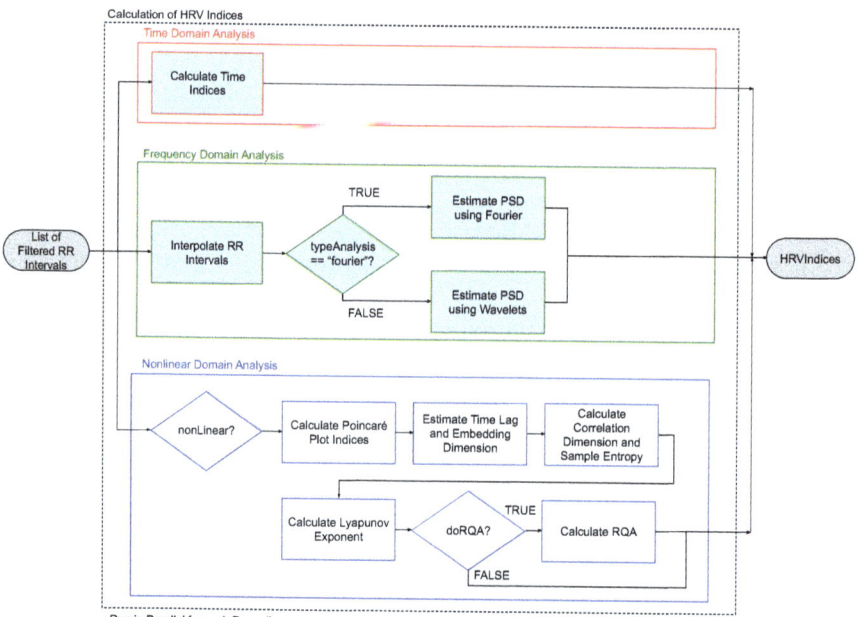

Fig. 8.4 Workflow with the operations performed by *RHRVEasy* during the calculation of the HRV indices

8.3 Statistical Analysis

The frequency index calculation starts by interpolating the heart rate (*InterpolateNIHR()*), defaulting to 4 Hz using linear interpolation [1]. Then a frequency analysis is created (*CreateFreqAnalysis()*). The *typeAnalysis* parameter determines whether the Fourier transform (default option) or the wavelet transform is used (see Sect. 4.2). The default *RHRVEasy* spectral bands are the ones used by the function *CalculateEnergyInPSDBands*: $ULFmin = 0$, $ULFmax = 0.03$, $VLFmin = 0.03$, $VLFmax = 0.05$, $LFmin = 0.05$, $LFmax = 0.15$, $HFmin = 0.15$, and $HFmax = 0.4$. When the *typeAnalysis* is set to "*wavelet*", the default mother wavelet is the extremal phase Daubechies of width 4, and the *bandtolerance* parameter used to estimate the limits of the frequency bands is set to 0.01.

To keep computing time within reasonable limits, when the *RHRVEasy* function is called using the default parameters, it only calculates the time and frequency indices. Later in this chapter, we shall see how the calculation of the nonlinear indices is tackled.

8.3 Statistical Analysis

After calculating the HRV indices, the statistical analysis commences (see Fig. 8.5). This analysis begins by constructing an ANOVA model for each index. The normality of the residuals is then evaluated using the Shapiro-Wilk test. If the test indicates that an index is not normally distributed, the ANOVA model is discarded. In these cases, the nonparametric Kruskal-Wallis test is employed to identify potential differences between the populations for that specific index.

Due to the multiple statistical tests (one per HRV index), a correction for the significance level should be applied. The default correction method is Bonferroni ("*bonferroni*"), which divides the p-values by the number of tests performed [2]. The significance level used by default when reporting the results is 0.05. This value can be overwritten in the *RHRVEasy* function call via the parameter *significance*.

The Bonferroni correction method is quite conservative, reducing the statistical power of the analysis and increasing the probability of Type II errors. The *correctionMethod* parameter permits selecting other significance level corrections. The supported values include "*holm*", "*hochberg*", "*hommel*", "*BH*" (Benjamini and

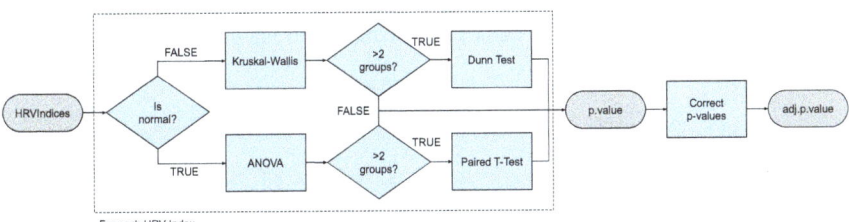

Fig. 8.5 Steps followed by *RHRVEasy* during the statistical analysis

Hochberg), *"fdr"* (false discovery rate or FDR), *"BY"* (Benjamini and Yekutieli), and *"none"*. The Holm, Hochberg, Hommel, and Bonferroni methods are designed to control the family-wise error rate [3], decreasing the chance of Type I errors. The Benjamini and Hochberg, Benjamini and Yekutieli, and FDR methods aim to keep the predicted fraction of erroneous findings under control; they have a higher statistical power, although they increase the chance of Type I errors [4]. If the *correctionMethod* is set to *"none"*, no correction is applied to the p-values. When a correction is applied, both the corrected and uncorrected p-values are stored and will be available at the end of the statistical analysis, as shown in Fig. 8.3.

8.4 Saving and Inspecting the Analysis Results

The *RHRVEasy* function returns a *RHRVEasyResult* object containing the HRV indices for each recording stored in a *data.frame* named *HRVIndices*. The *RHRVEasy* object also contains all the statistical test results for each HRV index (either based on ANOVA or Kruskal-Wallis depending on the normality of the index). These are stored in the *data.frame stats*, which also contains both the uncorrected and corrected p-values (see Fig. 8.3).

When the *RHRVEasy* object is printed in the console, the *RHRVEasy* object summarizes the information about the HRV indices that showed statistically significant differences between the populations (see Listing 8.4.1 for an illustrative example). This summary includes p-values and unadjusted confidence intervals. The calculation of these intervals utilizes t-test-based methods for normally distributed indices and bootstrapping [5] for non-normal indices. Although all the information regarding the indices that did not show statistically significant differences is available in the *RHRVEasy* object and may be inspected using R code, these indices are not reported when printing the object in the console.

If the *saveHRVIndicesInPath* argument is specified during the *RHRVEasy* function call, an Excel spreadsheet containing all the HRV indices for each recording is created in the specified path. It is also possible to save this information using the function *SaveHRVIndices*, which receives as arguments an *RHRVEasy* object and the path where the Excel file should be saved. By default, the name of the file that is generated is *experimental_group1_Vs_experimental_group2.xlsx*, being *experimental_group1* and *experimental_group2* the names of the groups being compared in the analysis. Other names can be indicated via the argument *filename*.

R listing 8.4.1

```
# Fragment of the output of a RHRVEasy object when
# displayed on the console. The text output format
# has been adapted to enhance its readability in the book.

## Significant differences in SDNN (Kruskal-Wallis rank
## sum test, bonferroni p-value = 1.117154e-07):
##   chf's mean95% CI: (62.17953, 93.58125)
```

```
##                    [Bootstrap CI without adjustment]
##    normal's mean95% CI: (131.7343, 148.0405)
##                    [Bootstrap CI without adjustment]
##
## Significant differences in SDANN (Kruskal-Wallis rank
## sum test, bonferroni p-value = 3.799696e-07):
##    chf's mean95% CI: (48.07725, 80.67042)
##                    [Bootstrap CI without adjustment]
##    normal's mean95% CI: (121.2193, 137.836)
##                    [Bootstrap CI without adjustment]
```

8.5 A First Example

To illustrate the functionalities of *RHRVEasy*, we will utilize several publicly available databases. In this first example, we will use a database containing recordings from healthy subjects and others consisting of several recordings from patients diagnosed with congestive heart failure (CHF). The first one is the Normal Sinus Rhythm RR Interval Database [6] (hereinafter referred to as NSR_DB). It contains 54 long-term recordings (approximately 24 h each) from 30 men (aged 28–76) and 24 women (aged 58–73). The second is the Congestive Heart Failure RR Interval Database [6] (hereinafter referred to as CHF_DB). This database contains 29 long-term recordings (around 24 h each) from subjects aged 34–79. Eight men and two women are included, while the gender of the remaining subjects is unknown. Both databases originated from ECG recordings digitized at a sampling rate of 128 Hz. The beat annotations of both databases were obtained through automated analysis, followed by manual review and correction.

The GitHub repository for this book includes both databases after extracting the RR intervals from the beat annotations [7]. These databases are located within the directory *data/Chapter8*, organized into subfolders named *normal* and *chf*. Listing 8.5.1 contains variables that point to these directories and that will be used by the code in this chapter.

R listing 8.5.1

```
# Variables that contain the paths of the NSR_DB
# and the CHF_DB databases.
library(RHRV)
basePath <- "data/Chapter8"
NSR_DB <- file.path(basePath, "normal")
CHF_DB <- file.path(basePath, "chf")
```

If we want to calculate all the time- and frequency-domain indices for the 54 recordings of the NSR_DD and for the 29 of the CHF_DB, and then perform a statistical comparison with significance level correction, we only have to write a single line of R code, as shown in Listing 8.5.2.

R listing 8.5.2

```
# Analysis of the NSR_DB and the CHF_DB databases
# using RHRVEasy.
easyAnalysis <- RHRVEasy(folders = c(NSR_DB, CHF_DB))
```

And that is it! To inspect the analysis results, you can either display the object in the console or access its slots *HRVIndices* and *stats*, as shown in Listing 8.5.3.

R listing 8.5.3

```
# Print output adapted for brevity
print(easyAnalysis)

   ## Significant differences in SDNN (Kruskal-Wallis rank
   ## sum test, bonferroni p-value = 1.117154e-07):
   ##    chf's mean 95% CI: (62.7952, 93.19571)
   ##
   ##                      [Bootstrap CI without adjustment]
   ##    normal's mean 95% CI: (131.3271, 147.9712)
   ##
   ##                      [Bootstrap CI without adjustment]
   ##
   ## Significant differences in SDANN (Kruskal-Wallis rank
   ## sum test, bonferroni p-value = 3.799696e-07):
   ##    chf's mean 95% CI: (47.24639, 81.30971)
   ##
   ##                      [Bootstrap CI without adjustment]
   ##    normal's mean 95% CI: (121.6326, 137.9365)
   ##
   ##                      [Bootstrap CI without adjustment]
   ##
   ## Significant differences in SDNNIDX (Kruskal-Wallis rank
   ## sum test, bonferroni p-value = 0.01426098):
   ##    chf's mean 95% CI: (29.1096, 47.97392)
   ##
   ##                      [Bootstrap CI without adjustment]
   ....

# The analysis results can also be accessed using the
# slots $HRVIndices and $stats. For HRVIndices, show only
# the first 6 columns
head(easyAnalysis$HRVIndices[, 1:6])

   ## # A tibble: 6 x 6
   ##    file                group  SDNN SDANN SDNNIDX pNN50
   ##    <chr>               <fct>  <dbl> <dbl>  <dbl> <dbl>
   ## 1 chf201_rr_secs.txt chf    75.5  52.9    49.6  2.03
   ## 2 chf202_rr_secs.txt chf    88.5  75.8    39.6  6.13
   ## 3 chf203_rr_secs.txt chf    38.8  30.9    21.7  1.20
   ## 4 chf204_rr_secs.txt chf    55.1  39.1    36.0  4.84
   ## 5 chf205_rr_secs.txt chf    34.9  26.1    19.5  1.97
   ## 6 chf206_rr_secs.txt chf    41.2  34.9    14.8  2.02

head(easyAnalysis$stats)

   ## # A tibble: 6 x 4
   ##    HRVIndex method                              p.value adj.p.value
   ##    <chr>    <chr>                                 <dbl>       <dbl>
```

```
## 1 SDNN     Kruskal-Wallis rank sum test 0.00000000798 0.000000112
## 2 SDANN    Kruskal-Wallis rank sum test 0.0000000271  0.000000380
## 3 SDNNIDX  Kruskal-Wallis rank sum test 0.00102       0.0143
## 4 pNN50    Kruskal-Wallis rank sum test 0.774         1
## 5 SDSD     Kruskal-Wallis rank sum test 0.0891        1
## 6 rMSSD    Kruskal-Wallis rank sum test 0.0891        1
```

The analysis reveals statistically significant differences in eight HRV indices: six time-domain indices (SDNN, SDANN, SDNNIDX, IRRR, TINN, and HRVi) and two frequency-domain indices (ULF and VLF). It is known that patients with CHF present marked differences in VLF, and lesser differences in LF, but no differences in the HF band [8]. The results obtained for the temporal indices are also consistent with what has been previously reported in the scientific literature [9, 10].

8.5.1 Changing the Main Default Parameters

By default, the frequency analysis is performed using the Fourier transform. If we pass the value *"wavelet"* to the *typeAnalysis* parameter, the analysis will be carried out using the wavelet transform. In this case, the same two frequency bands as in the Fourier-based analysis (ULF and VLF) present statistically significant differences (see Listing 8.5.4):

R listing 8.5.4

```
# Analysis of the NSR_DB and the CHF_DB databases using
# the wavelet transform for frequency indices calculation.
easyAnalysisWavelet <- RHRVEasy(folders = c(NSR_DB, CHF_DB),
                                typeAnalysis = "wavelet")
```

To save the results in an Excel file, we can specify the argument *saveHRVIndicesInPath* in the *RHRVEasy* function call, as illustrated by Listing 8.5.5.

R listing 8.5.5

```
# When specifying a value for the argument saveHRVIndicesInPath,
# an Excel file with the HRV indices is created at the path
# specified by the argument
spreadsheetPath <- basePath
easyAnalysis <- RHRVEasy(folders = c(NSR_DB, CHF_DB),
                         saveHRVIndicesInPath = spreadsheetPath)
```

This spreadsheet can also be generated from the object returned by *RHRVEasy* by calling the function *SaveHRVIndices* (see Listing 8.5.6).

R listing 8.5.6

```
# The HRV indices contained in a RHRVEasy object called
# easyAnalysis are saved in an Excel file at the specified path
SaveHRVIndices(easyAnalysis,
               saveHRVIndicesInPath = spreadsheetPath)
```

If we want to use another correction method different from the default one (*"bonferroni"*), it can be specified through the parameter *correctionMethod*, as shown in Listing 8.5.7.

R listing 8.5.7

```
# The false discovery rate ("fdr") method is employed
# to adjust for significance level
easyAnalysisFDR <- RHRVEasy(folders = c(NSR_DB, CHF_DB),
                            correctionMethod = "fdr")
print(easyAnalysisFDR)
```

With the FDR correction method, the LF index also presents statistically significant differences between both databases. According to the literature, this band presents differences between patients with CHF and healthy subjects, although more subtle than in the case of the VLF band [8].

The code from Listing 8.5.7 had to recalculate all the time and frequency indices of both databases and then apply the new correction of the significance level. If we already have the results of an *RHRVEasyResult* object, we can apply a different correction method using the function *RHRVEasyStats*. This function receives as arguments an *RHRVEasyResult* object and the new correction method. This function does not recalculate the indices, but it just applies the new correction method to the uncorrected p-values stored in the object passed as an argument. Listing 8.5.8 applies the *"fdr"* correction method on the object *easyAnalysis* (which used the *"bonferroni"* correction), and shows in the console a comparison of the p-values without correction, and applying both corrections.

R listing 8.5.8

```
# The "fdr" correction method is applied over the p-values
# contained in the object easyAnalysis, (which was created
# using the "bonferroni" correction)
easyAnalysisFDR <- RHRVEasyStats(easyAnalysis,
                                 correctionMethod = "fdr")
pValues <- merge(
  easyAnalysis$stats,
  easyAnalysisFDR$stats,
  by = setdiff(names(easyAnalysis$stats), "adj.p.value"),
  suffixes = c(".bonf", ".fdr")
)
pValues <- pValues[, c("HRVIndex", "p.value",
                       "adj.p.value.bonf", "adj.p.value.fdr")]
```

8.6 Comparing More than Two Experimental Groups

```
pValues[, 2:4] <- round(pValues[, 2:4], 4)
#A comparison of the p-values without correction,
# and applying both corrections is shown in the console
print(head(pValues))
##   HRVIndex p.value adj.p.value.bonf adj.p.value.fdr
## 1       HF  0.5601           1.0000          0.6032
## 2     HRVi  0.0000           0.0000          0.0000
## 3     IRRR  0.0000           0.0000          0.0000
## 4       LF  0.0165           0.2312          0.0257
## 5    MADRR  0.0632           0.8848          0.0885
## 6    pNN50  0.7745           1.0000          0.7745
```

Note that, as expected, the p-values are larger when using Bonferroni than when using FDR and that the latter p-values are also larger than those obtained without applying any correction method.

8.6 Comparing More than Two Experimental Groups

HRV studies may sometimes involve more than two populations. For instance, a study might compare two different treatments for a cardiac condition with a third control group. In these cases, even if a global test such as ANOVA or Kruskal-Wallis reveals significant differences among the groups, these differences might not exist between every possible pair of groups. Post hoc tests are employed to pinpoint which specific pairs of groups actually differ from each other. The selection of the appropriate post hoc test depends on the initial test used: paired t-tests follow ANOVA, while Dunn's test [11] is typically used after Kruskal-Wallis. The significance level must also be adjusted to account for the multiple comparisons performed during the post hoc tests. Note that both the results of the global tests (either ANOVA or Kruskal-Wallis) and the post hoc tests (paired t-tests or Dunn's test) are available in the *RHRVEasyResult* object, as illustrated in Fig. 8.3. The latter are accessible as a nested *data.frame* for those global tests that were significant.

The *RHRVEasy* function requires a list of folders as its only mandatory argument. This list must contain at least two folders, but there is no upper limit on the number of folders it can contain. Therefore, to analyze three or more experimental groups, we just include the folders containing their recordings in this list.

To demonstrate an analysis of multiple experimental groups, another publicly available database will be used: the RR interval time series from healthy subjects [6, 12]. This database contains RR time series of 147 healthy individuals: 72 males and 67 females and 8 individuals for whom gender data is unavailable. Of the 147 individuals, 72 were under 1 year old, while 9 were over 18 years old. To focus on a population similar to that from the NSR_DB, we will analyze the recordings from adult individuals (recordings: *"000.txt"*, *"003.txt"*, *"005.txt"*, *"006.txt"*, *"007.txt"*, *"008.txt"*, *"009.txt"*, *"010.txt"*, and *"013.txt"*). Listing 8.6.1 shows the code to download these files, although they can also be found in the GitHub repository that contains the code used in this book [7], under the folder *data/Chapter8/healthy_ms*.

R listing 8.6.1

```r
# paste0 is just used to split the long url
orig <- paste0(
  "https://www.physionet.org/files/",
  "rr-interval-healthy-subjects/1.0.0/"
)
dest <- file.path(basePath, "healthy_ms")
# Create folder dest if not exists
if (!dir.exists(dest)) {
  dir.create(dest)
}
files <- c("000.txt", "003.txt", "005.txt", "006.txt",
          "007.txt", "008.txt", "009.txt", "010.txt",
          "013.txt")

for (file in files) {
  download.file(paste0(orig, file), file.path(dest, file))
}
```

Inspecting any of these records reveals that the RR intervals are stored in milliseconds. To enable joint analysis with NSR_DB and CHF_DB datasets, which are stored in seconds, we will convert them to seconds as shown in Listing 8.6.2. The resulting files can also be found in the GitHub repository, under the folder *data/Chapter8/healthy*.

R listing 8.6.2

```r
orig <- file.path(basePath, "healthy_ms")
HEALTHY_DB <- file.path(basePath, "healthy")
# Create folder dest if not exists
if (!dir.exists(HEALTHY_DB)) {
  dir.create(HEALTHY_DB)
}
for (file in list.files(orig)) {
  rrs_ms <- read.table(file.path(orig, file))
  rrs_ms <- rrs_ms$V1
  # Recording 008 contains an @
  rrs_ms <- as.numeric(rrs_ms[rrs_ms != "@"])
  # Convert from milliseconds to seconds
  write.table(rrs_ms / 1000, file.path(HEALTHY_DB, file),
              row.names = FALSE, col.names = FALSE)
}
```

Listing 8.6.3 shows the code for analyzing the three databases and presenting the statistical analysis results. Note that this output differs from that in Listing 8.5.3, as it now includes post hoc tests. The same significant indices found in Listing 8.5.3 remain significant, with a new index, MADRR, added to the list. Post hoc tests

8.6 Comparing More than Two Experimental Groups

reveal differences between NSR_DB and CHF_DB (in seven HRV indices; note that the SDNNIDX comparison is not significant in the post hoc test, probably due to multiple comparisons) and between HEALTHY_DB and CHF_DB (in seven other HRV indices). Notably, there are no differences found between NSR_DB and HEALTHY_DB.

R listing 8.6.3

```
easyAnalysis3 <-
  RHRVEasy(folders = c(NSR_DB, CHF_DB, HEALTHY_DB))
# Print output adapted for brevity
print(easyAnalysis3)

## Significant differences in SDNN (Kruskal-Wallis rank
## sum test, bonferroni p-value = 3.543622e-07):
##   Significant differences in the post-hoc tests
##   (Dunn's all-pairs test + bonferroni-p-value adjustment):
##       group1 group2 adj.p.value
##     1 healthy chf   0.00799
##     2 normal  chf   0.000000282
##   ----------------------------
##     chf's mean 95% CI: (62.29769, 92.93011)
##                     [Bootstrap CI without adjustment]
##     healthy's mean 95% CI: (123.1111, 156.1362)
##                     [Bootstrap CI without adjustment]
##     normal's mean 95% CI: (131.4674, 148.1886)
##                     [Bootstrap CI without adjustment]
##
....
```

As shown in Listing 8.6.4, the inspection of the *$stats* slot reveals the existence of a new column *pairwise* storing the results of the post hoc comparisons.

R listing 8.6.4

```
print(head(easyAnalysis3$stats))

## # A tibble: 6 x 5
##   HRVIndex method                                p.value      adj.p.value pairwise
##   <chr>    <chr>                                 <dbl>        <dbl>       <list>
## 1 SDNN     Kruskal-Wallis rank sum test 0.0000000253 0.000000354 <tibble>
## 2 SDANN    Kruskal-Wallis rank sum test 0.0000000961 0.00000135  <tibble>
## 3 SDNNIDX  Kruskal-Wallis rank sum test 0.0000760    0.00106     <tibble>
## 4 pNN50    Kruskal-Wallis rank sum test 0.0186       0.260       <NULL>
## 5 SDSD     Kruskal-Wallis rank sum test 0.0301       0.421       <NULL>
## 6 rMSSD    Kruskal-Wallis rank sum test 0.0301       0.421       <NULL>

# Let's print the results of the post-hoc tests for
# the SDNN index
easyAnalysis3$stats[easyAnalysis3$stats$HRVIndex == "SDNN", ]$pairwise[[1]]

## # A tibble: 3 x 6
##   HRVIndex group1 group2 method                     p.value adj.p.value
```

```
##   <chr>      <chr>  <chr>   <chr>                            <dbl>         <dbl>
## 1 SDNN    healthy  chf     Dunn's all-pairs test 0.000296          0.00799
## 2 SDNN    normal   chf     Dunn's all-pairs test 0.0000000104 0.000000282
## 3 SDNN    normal   healthy Dunn's all-pairs test 0.861              1
```

8.7 Nonlinear Index Calculation

Calculating certain nonlinear HRV indices automatically presents a challenge. Ideally, this process relies on input from the user, who on several occasions has to make decisions about the parameters to be used in the analysis of each recording based on the visual inspection of plots. Additionally, calculating these indices, particularly those based on recurrence quantification analysis (RQA), demands significant computational resources. While computing all time- and frequency-domain indices for a few dozen recordings in a typical HRV analysis might only take a few seconds, calculating the nonlinear indices could take up to several hours. This is why the *RHRVEasy* function defaults the Boolean parameters *nonLinear* and *doRQA* to *FALSE*. Only if these parameters are changed will the nonlinear analysis and RQA be executed (see Fig. 8.4 for an overview of the steps followed).

The nonlinear analysis starts by initializing the necessary RHRV structure (*CreateNonLinearAnalysis()*). The analysis begins by calculating the indices derived from the Poincaré plot (SD_1 and SD_2 indices). These are the only ones guaranteed to be calculated for each recording, as the computation of the remaining nonlinear indices may run into different failures, as explained below.

Then the time delay and embedding dimension are calculated. This process begins with computing the autocorrelation function of the RR time series to determine the time lag (*CalculateTimeLag()*). *RHRVEasy* first searches for the first minimum in the autocorrelation function. If no minimum is found, the first value that decays to $1/e$ is searched for. If none of these criteria are met, the average mutual information is used for time delay estimation. It again searches for the first minimum or the first point decaying to the peak value/e. If all these attempts to determine the time lag fail, a heuristic value of 30 is set, as this value has been shown through experimentation to sufficiently minimize autocorrelation. Once the time lag has been derived, the embedding dimension is then estimated (*CalculateEmbeddingDim()*).

Once the time lag and the embedding dimension have been computed, the calculation of the correlation dimension can begin (*CalculateCorrDim()*, *EstimateCorrDim()*). This is typically done by plotting the correlation sum $C(m, r)$ against the radius r for various embedding dimensions m and looking for the linear scaling region (see Fig. 5.10). This linear region should be preceded in the plot by the oscillatory region and the noise regime region and followed by the macroscopic regime (see Fig. 5.10 and Sect. 5.2.4). Given the goal of *RHRVEasy*, some form of automation of the identification of these four regions is necessary. *RHRVEasy* fits a piecewise linear regression with four regions to the plot of the local scaling

8.7 Nonlinear Index Calculation

exponents versus the radius. The flattest segment in this plot is chosen as the scaling region. The sample entropy is calculated using this same region (*EstimateSampleEntropy()*). To obtain a more robust estimate of both statistics, the procedure is repeated three times using embedding dimensions from *embeddingDimension* up to *embeddingDimension* + 2. The final value for each statistic is the average of the three estimates.

The estimation of the Lyapunov exponent follows a strategy similar to that of the correlation dimension. The exponential divergence of initially close trajectories in phase space is plotted against time. This plot is expected to exhibit two regions, with the first displaying a constant slope (see Fig. 5.12). A piecewise linear regression is then fitted to the plot, separating it into two regions. The Lyapunov exponent is estimated (*EstimateMaxLyapunov()*) from the slope of the first region. This process is repeated for the same embedding dimensions used for the correlation dimension calculation, and the average of these values is the final estimate.

It is important to note that estimating the Lyapunov exponent requires defining a radius to identify close trajectories. In Sect. 5.4.6, the *radius* parameter was selected based on the correlation sum. This heuristic relied on interpreting the correlation sum $C(m, r)$ as the average probability of finding a neighbor in a ball of radius r when using an embedding dimension m. *RHRVEasy* automates this process by selecting a small radius r_{small} fulfilling $\overline{C}(r_{small}) \approx 10^{-3}$, where $\overline{C}(r)$ represents the average correlation sum $C(m, r)$ across all dimensions.

Finally, the RQA analysis is performed using the same r_{small} that was employed for the Lyapunov calculations (*RQA()*). All the nonlinear indices computed are stored in the *HRVIndices data.frame*, along the temporal and frequency indices.

Calculating the nonlinear indices, except for those derived from the Poincaré plot, can encounter certain situations that prevent their computation. The piecewise linear regressions used in the calculations of the correlation dimension, the sample entropy, and the maximal Lyapunov exponent may fail to identify the expected number of linear regions. Furthermore, some RQA-derived indices involve division by quantities that might be zero for certain recordings. If a statistic cannot be computed due to any of these reasons, its value is set to NA (not available).

Listing 8.7.1 shows how to perform a complete analysis, including all nonlinear HRV indices, on the three datasets used in the previous section. It should be noted that this code will take several hours to run on a PC with medium-high computing power.

The Poincaré's SD_2 index and several RQA indices (maximal line lengths of vertical and horizontal lines, trapping time, divergence, and laminarity) present statistically significant differences. Specifically, post hoc test results indicate significant differences between the pairs NSR_DB-CHF_DB and HEALTHY_DB-CHF_DB, while no significant differences are observed between NSR_DB and HEALTHY_DB. Finding differences in SD_2, which characterizes long-term variability, and not in SD_1, which characterizes short-term variability, is consistent with having found differences in the lower-frequency bands of the spectral analysis, but not in the higher ones. On the other hand, the alterations found in RQA are consistent with the findings of other authors who have shown that the indices derived from RQA tend to present the differences between healthy subjects and patients with chronic congestive failure [13].

R listing 8.7.1

```
# Analysis of the three databases calculating all
# the non-linear indices
fullAnalysis <- RHRVEasy(
  folders = c(NSR_DB, CHF_DB, HEALTHY_DB),
  nonLinear = TRUE,
  doRQA = TRUE,
)
# Print output adapted for brevity
print(fullAnalysis)
```

8.8 Parallelization of the Calculations

The computation time of the HRV indices can be high for long-term recordings or when the experimental groups contain a large number of subjects. This is especially true for nonlinear indices. For example, analyzing one of the 24-h RR recordings used in this chapter takes about 1 s on a midrange computer, considering only time- and frequency-domain analyses. However, calculating nonlinear indices for the same recording can take up to 6 min.

We have parallelized the processing using the *foreach* R package to speed up the calculations by taking advantage of several CPU cores. This enables the simultaneous processing of multiple recordings at a time, reducing analysis time. The degree of parallelization can be controlled by the *nJobs* parameter within the *RHRVEasy* function, which defaults to using a single CPU core. Setting *nJobs* ≤ 0 uses all cores available in the system. In multicore systems, the value of this parameter can be changed to take advantage of all of the available cores. Listing 8.8.1 repeats the analysis of Listing 8.7.1 using eight cores. In a system with this number of cores, this may lower the execution time by a factor of about 8 (in our tests, it reduced the calculation time to approximately 1 h).

R listing 8.8.1

```
#Parallelized analysis of the four databases
#using 8 cores
fullAnalysis <- RHRVEasy(
  folders = c(NSR_DB, CHF_DB, HEALTHY_DB),
  nonLinear = TRUE,
  doRQA = TRUE,
  nJobs = 8
)
print(fullAnalysis)
```

8.9 Conclusions

RHRVEasy streamlines HRV analysis by automating the calculation of a wide range of HRV indices and their subsequent statistical analysis. This automation significantly reduces time and effort for researchers. Without such a tool, researchers would have to repeat the same analysis steps for each recording across every population studied.

However, obtaining this complete automation necessitates certain trade-offs and design decisions that limit the analysis type and scope. For instance, *RHRVEasy* currently does not support comparisons between indices calculated in different windows of the same recording or across a set of recordings. We aim to address this limitation in future versions.

Another limitation is that the optimal calculation of some nonlinear indices requires a visual inspection of plots to identify the most appropriate linear regions for calculating the nonlinear statistics. Therefore, the nonlinear indices estimated by *RHRVEasy* should be considered initial approximations suitable only for exploratory analysis. If these indices reveal significant results, we recommend visually inspecting the plots for each recording to select the most suitable parameters for calculating the nonlinear statistics.

It is also important to consider that the default parameters used for calculating the indices might not be suitable for all populations. These parameters are appropriate for adult humans at rest or during low-intensity exercise and may not be optimal for all populations or activities. If the analysis involves subjects who are doing sports [14], children [15], or animals [16, 17], it is necessary to adjust the parameters, especially those related to the spectral analysis.

Despite these limitations, the authors believe that *RHRVEasy* will become a valuable tool for researchers in the field of HRV, facilitating their work, saving time, and thus contributing to more rapid progress in the field.

References

1. K.K. Kim, J.S. Kim, Y.G. Lim, K.S. Park, The effect of missing RR-interval data on heart rate variability analysis in the frequency domain. Physiol. Meas. **30**(10), 1039 (2009). [Online]. Available: https://doi.org/10.1088/0967-3334/30/10/005
2. R.A. Armstrong, When to use the bonferroni correction. Ophthalmic Physiol. Opt. **34**(5), 502–508 (2014). [Online]. Available: https://doi.org/10.1111/j.1475-1313.2010.00815.x
3. J. Stange, T. Dickhaus, A. Navarro, D. Schunk, Multiplicity-and dependency-adjusted p-values for control of the family-wise error rate. Stat. Probab. Lett. **111**, 32–40 (2016). [Online]. Available: https://doi.org/10.1016/j.spl.2016.01.005
4. M.E. Glickman, S.R. Rao, M.R. Schultz, False discovery rate control is a recommended alternative to bonferroni-type adjustments in health studies. J. Clin. Epidemiol. **67**(8), 850–857 (2014). [Online]. Available: https://doi.org/10.1016/j.jclinepi.2014.03.012
5. A.C. Davison, D.V. Hinkley, *Bootstrap Methods and Their Application* (Cambridge University Press, Cambridge, 1997), no. 1. [Online]. Available: https://doi.org/10.1017/CBO9780511802843

6. A.L. Goldberger, L.A. Amaral, L. Glass, J.M. Hausdorff, P.C. Ivanov, R.G. Mark, J.E. Mietus, G.B. Moody, C.-K. Peng, H.E. Stanley, Physiobank, physiotoolkit, and physionet: components of a new research resource for complex physiologic signals. Circulation **101**(23), e215–e220 (2000). [Online]. Available: https://doi.org/10.1161/01.cir.101.23.e215
7. C.A. García, A. Otero, X.A. Vila, M.J. Lado, L. Rodríguez-Liñares, J.M. Presedo, A.J. Méndez, Code for the book 'Heart Rate Variability Analysis with the R package RHRV' (Springer's use R!). https://github.com/RHRV-team/RHRVBook (2024)
8. M. Hadase, A. Azuma, K. Zen, S. Asada, T. Kawasaki, T. Kamitani, S. Kawasaki, H. Sugihara, H. Matsubara, Very low frequency power of heart rate variability is a powerful predictor of clinical prognosis in patients with congestive heart failure. Circ. J. **68**(4), 343–347 (2004). [Online]. Available: https://doi.org/10.1253/circj.68.343
9. P. Melillo, N. De Luca, M. Bracale, L. Pecchia, Classification tree for risk assessment in patients suffering from congestive heart failure via long-term heart rate variability. IEEE J. Biomed. Health. Inf. **17**(3), 727–733 (2013). [Online]. Available: https://doi.org/10.1109/jbhi.2013.2244902
10. A. Musialik-Łydka, B. Średniawa, S. Pasyk, Heart rate variability in heart failure. Kardiologia Polska (Polish Heart J.) **58**(1), 14–16 (2003)
11. A. Dinno, Nonparametric pairwise multiple comparisons in independent groups using dunn's test. Stata J. **15**(1), 292–300 (2015). [Online]. Available: https://doi.org/10.1177/1536867X150150011
12. I. Irurzun, L. Garavaglia, M. Defeo, J.T. Mailland Rr interval time series from healthy subjects. Last accessed: 13 May 2024. [Online]. Available: https://physionet.org/content/rr-interval-healthy-subjects/1.0.0/ (2021)
13. N. Wessel, N. Marwan, U. Meyerfeldt, A. Schirdewan, J. Kurths, Recurrence quantification analysis to characterise the heart rate variability before the onset of ventricular tachycardia, in *International Symposium on Medical Data Analysis* (Springer, Berlin, 2001), pp. 295–301
14. L. Schmitt, J. Regnard, M. Desmarets, F. Mauny, L. Mourot, J.-P. Fouillot, N. Coulmy, G. Millet, Fatigue shifts and scatters heart rate variability in elite endurance athletes. PloS One **8**(8), e71588 (2013). [Online]. Available: https://doi.org/10.1371/journal.pone.0071588
15. A. Martín-Montero, G.C. Gutiérrez-Tobal, L. Kheirandish-Gozal, J. Jiménez-García, D. Álvarez, F. Del Campo, D. Gozal, R. Hornero, Heart rate variability spectrum characteristics in children with sleep apnea. Pediatr. Res. **89**(7), 1771–1779 (2021). [Online]. Available: https://doi.org/10.1038/s41390-020-01138-2
16. M. Kuwahara, K.-i. Yayou, K. Ishii, S.-i. Hashimoto, H. Tsubone, S. Sugano, Power spectral analysis of heart rate variability as a new method for assessing autonomic activity in the rat. J. Electrocardiol. **27**(4), 333–337 (1994). [Online]. Available: https://doi.org/10.1016/s0022-0736(05)80272-9
17. M. Kuwahara, S.-i. Hashimoto, K. Ishii, Y. Yagi, T. Hada, A. Hiraga, M. Kai, K. Kubo, H. Oki, H. Tsubone, et al., Assessment of autonomic nervous function by power spectral analysis of heart rate variability in the horse. J. Auton. Nerv. Syst. **60**(1–2), 43–48 (1996). [Online]. Available: https://doi.org/10.1016/0165-1838(96)00028-8

Appendix A
Installing RHRV

A.1 RHRV Installation

This guide assumes that the user has some basic knowledge of the R environment. If this is not your case, you can find a nice introduction to R in the R project homepage [1]. The R project homepage also provides an "R Installation and Administration" guide. Once you have download and installed R, you can install RHRV by typing the following:

```
> install.packages("RHRV")
```

You will be asked to select a CRAN mirror to download the package. The CRAN acronym refers to "The Comprehensive R Archive Network," a network of servers that store identical, up-to-date, versions of R packages [2].

You can also install RHRV by directly downloading it from the CRAN. Once the download has finished, open R, move to the directory where you have download it (by using the R command `setwd`), and type the following:

```
> install.packages("RHRV_XXX",repos=NULL)
```

Here, XXX is the version number of the library. To start using the library, you should load it by using the library command:

```
> library(RHRV)
```

Finally, you may want to install the development version of RHRV, available at R-Forge [3]. The development version contains the latest updates introduced into the RHRV package. However, it must be noted that this is an unstable version, and thus it may contain bugs. To install it, just type the following:

```
> install.packages("RHRV",
                   repos="http://R-Forge.R-project.org")
```

References

1. The R project for statistical computing. Last accessed 15 Apr 2024. [Online]. Available: http://www.r-project.org/
2. The comprehensive R archive network. Last accessed 15 Apr 2024. [Online]. Available: http://cran.r-project.org/
3. The RHRV repository. Last accessed 15 Apr 2024. [Online]. Available: http://r-forge.r-project.org/R/?group_id=919

Appendix B
How Do I Get a Series of RR Intervals from a Clinical/Biological Experiment?

The analysis of HRV is based on the availability of a time series of the beats's positions. Such time series is usually produced by an algorithm that operates over the ECG signal, but it could be provided by any algorithm that processes other physiological signals and that provide a time series of similar characteristics.

The main purpose of the HRV analysis is to investigate the influence of the ANS on the heart rate. Therefore, the onset of the P-wave is the appropriate fiducial point to be used [1]. However, a fiducial point related to the P-wave is difficult to obtain with enough accuracy because its poor signal to noise ratio: the P-wave usually has a very low amplitude and sometimes is missing. For such reason, the fiducial point is often taken inside the QRS complex, given that it is easier to determine a stable point inside this part of the beat cycle. The more prominent wave inside the QRS complex is usually the R wave; hence, some feature of this wave is usually taken as a fiducial point (position of maximum height, maximum slope, etc.). This can be done because the PR interval can be considered as relatively constant, and therefore, the RR intervals give us the same information as the PP intervals.

HRV can be applied over the results of beat detection performed on other physiological parameters (blood pressure, SaO2, etc.). But the stability of the fiducial point derived from them is usually worse than if we use the ECG signal, and such lack of stability could distort the HRV analysis. For sake of brevity, we will concentrate on the acquisition of a time series for the HRV analysis from the ECG. A software capable of doing this is called a QRS detector, because it always takes the fiducial point inside the QRS complex.

B.1 QRS Detectors

A QRS detector must take into account that there are a very large number of QRS morphologies. The morphology could change not only among patients but also in

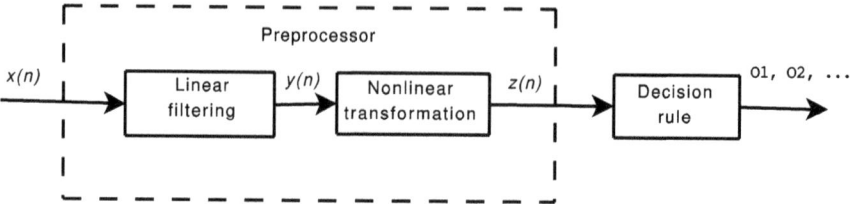

Fig. B.1 Block diagram of a commonly used QRS detector structure. The input is the ECG signal and the output is a series of occurrence times of the detected QRS complexes

time within the same patient. Furthermore, the detector must not only take into account certain rhythms, but it must also treat the next possible event as if it could occur at almost any time after the most recently detected beat.

QRS detection is a difficult task, not only because of the physiological variability of the QRS complexes but also because of the various types of noise that could be present in the ECG signal. Noise sources include muscle noise, artifacts due to electrode motion, power-line interference, baseline wander, and T waves with high-frequency characteristics similar to QRS complexes [2].

Any QRS detector that operates over one ECG channel could, in general terms, be described by the block diagram presented in Fig. B.1 [3].

- The linear filter is designed to have a bandpass behavior, so that the spectral content of the QRS complex is preserved, while unwanted ECG components such as the P and T waves are suppressed. The central frequency of the filter varies from 10 to 25 Hz and the bandwidth from 5 to 10 Hz. It is not relevant that this filter distorts the ECG because its purpose is to improve the SNR for the QRS detection.
- The nonlinear transformation further enhances the QRS complex in relation to the background noise and transforms each QRS complex into a single positive peak better suited for threshold detection.
- The decision rule takes the output of the preprocessor (the union of the linear filter and the nonlinear transformation) and performs a test on whether a QRS complex is present or not.

QRS detection strategies can be classified into two categories: single-channel and multichannel detectors [4]. The single-channel detector locates the QRS based on information collected from one ECG electrode. The multichannel detector relies on information from more than one electrode pair. The majority of QRS detection algorithms in the literature describe single-channel detectors.

A multichannel detector can lead to improved performance over a single-channel detector. If a channel has a very good signal quality, then a multichannel detector does not give improved performance. The additional information supplied by the extra channels is redundant. Under normal conditions, however, the signal quality varies in time and from channel to channel because of noise and artifacts. This results in different amounts of information in different channels. But noise and

B How Do I Get a Series of RR Intervals from a Clinical/Biological Experiment?   181

artifacts are less likely to occur on all channels at the same time. More channels of data provide a better chance to gather the information necessary to detect all the QRS complexes.

In general, a multichannel QRS detector should be as good or better than any single-channel QRS detector. But the difficulties related to designing a multichannel QRS detector that takes into consideration all the information provided by the additional channels makes this kind of QRS detectors complex and less common. Multichannel QRS detectors use one of the following three architectures to fusion data from different channels:

- Channel switching: a single channel is chosen (in terms of signal quality) from the available channels and QRS detection is done on one channel.
- Detection combination: single-channel detection is run on all channels separately. Detection marks are lately combined into a single file, deleting redundant ones.
- Channel combination: the signals from some or all of the channels are combined and then, single-channel detection is done on the combined channel.

B.2 Signal Conditioning

After applying a QRS detector, we obtain an unevenly set of events associated with the time occurrence of beats.

The easiest way to represent the RR intervals is through the interval tachogram, in which the events, occurring at $t_0, t_1, \ldots t_M$, are transformed into a discrete-time signal consisting of the successive intervals, i.e., the RR intervals:

$$d_{IT}(k) = t_k - t_{k-1} \quad k = 1, \ldots M \tag{B.1}$$

The interval tachogram is the heart rhythm representation upon which time-domain measures are usually computed, such as *SDNN, SDANN, r-MSSD,* and *pNN50*

The inverse interval tachogram $d_{IIT}(k)$ is the alternative representation to the interval tachogram and is defined by

$$d_{IIT}(k) = 1/(t_k - t_{k-1}) \quad k = 1, \ldots M \tag{B.2}$$

which reflects instantaneous heart rate, usually in beats/minute.

Often, you may be interested only in intervals between normal (sinus) beats; these are usually called NN intervals. In such case, the first step is to classify each beat as normal or not normal and then to delete all intervals that are not between normal beats. The exclusion of non-normal RR intervals represents an important step in conditioning the series in order to make HRV analysis more reliable. Since manual edition of a 24-h Holter recording is extremely laborious, automated exclusion procedures have been developed in order to accomplish rejection of artifacts. A simple approach is to apply an exclusion criterion by which an RR interval is considered abnormal if it deviates more than a given percentage from the mean length of the preceding RR intervals. Other approaches based on filtering to suppress out of range points could be applied.

The main problem when using either $d_{IT}(k)$ or $d_{IIT}(k)$ is that these signals are not indexed by a sample number, as is common in evenly sampled discrete-time signals. Consequently, the power spectral analysis of these two signals cannot be expressed in cycles per second. Transformation of the tachogram signals into evenly sampled time-domain signals is essential to obtain a spectral description in hertz, so prior to applying HRV spectral analysis techniques, we usually interpolate such signals to obtain an evenly sampled discrete-time signal using linear, spline interpolation or the integrated pulse frequency modulator (IPFM) [1].

B.3 Creating a WFDB Compatible Record

Many of the possible sources of RR intervals come from information extracted from records in WFDB format [5]. This is because it is a common output format in most of the QRS detectors. If the user is not going to work with WFDB formatted files, the installation of the WFDB package is not required for the proper functioning of RHRV. The WFDB package contains a large collection of specialized functions for processing and manipulating PhysioNet's databases [6] that are always in WFDB format.

The current version of WFDB software package may be downloaded in source form, as a compressed archive from [7]. If your system is a GNU/Linux, first, you must install missing prerequisites. These include gcc (the GNU Compiler Collection), related software development tools such as make, the libcurl library (if NETFILES support is desired), the XView libraries (needed for WAVE only), and X11 (needed by XView). All of these components are free (open-source) software available for all popular platforms, including GNU/Linux, Mac OS X, and Unix. On a Debian-based distribution, you would meet the prerequisites if you run this command as root:

```
apt-get install gcc libcurl4-openssl-dev libexpat1-dev
```

After that, you must download the file wfdb.tar.gz and unpack it:

```
tar xfvz wfdb.tar.gz
```

As a last step, you must configure and install the application. Enter as root on the directory created by `tar` and type the following commands (where m and n identify the version of the wfdb package):

```
cd wfdb-10.m.n
./configure
make install
make check
```

B   How Do I Get a Series of RR Intervals from a Clinical/Biological Experiment?   183

There are many ways to create a WFDB-compatible record. Here is an easy way to do so:

1. If the ECG is still in analog format, digitize it.
2. Write the samples into a file in text form, as a column of decimal numbers. If you have digitized more than one signal, use a separate column for each signal.
3. Read the documentation of `wrsamp` to see how to prepare a binary signal file and a header file from the text file [7]. Typically, you will need to use a command such as the following:

```
wrsamp -F 128 -G 102.4 -i ecg01.txt -o ecg01 0
```

This example reads a text file named `ecg01.txt` and creates the files needed for a record named `ecg01`, namely, a signal file named `ecg01.dat` and a header file named `ecg01.hea`. The arguments of -F and -G specify that the signal was sampled at 128 Hz and that the signal was amplified in such a way that a step of 1 mV would appear as sample values that differ by 102.4 units. The final argument (0) indicates that the leftmost column in the input (column 0) contains the data.

B.4   Creating a Beat Annotation File

At this point, you should have a PhysioBank-compatible record, including .dat and .hea files. Before you can begin studying RR intervals, however, you will need to know the exact locations (times of occurrence) of the QRS complexes in the ECG. In PhysioBank, this information is encoded in beat annotation files. Annotation files have names that begin with the record name and end with an *annotator name* (such as qrs, ecg, ann, or atr; other names are possible, though less commonly used). For example, a beat annotation file for record `ecg01` might have a name such as `ecg01.ann`.

If you do not have a beat annotation file for your record, there are several ways to create one using the PhysioToolkit software:

- Use `sqrs`, a good, fast, and simple QRS detector.
- Use `wqrs`, a reasonably fast QRS detector that generally works better than sqrs.
- Use `gqrs`, a very good and fast QRS detector that generally performs better than the others listed here. Like the others, gqrs is optimized for use with adult human ECGs; unlike the others, gqrs can be configured easily for analysis of infant, pediatric, and nonhuman ECGs.
- Use `ecgpuwave`, a very good QRS detector that also locates the P- and T-waves and their boundaries.

An example of the use of the gqrs command is as follows:

```
gqrs -r 100 -s 1
```

In this example, we ask gqrs to process channel 1 of record 100. As a result, we obtain an annotation file named `100.qrs`.

You should always review the beat annotation file generated by any of these detectors; although all of them work well in most cases, there is wide variability among ECG recordings, and any QRS detector will make errors if the data quality is insufficient. There are, once again, several ways to do this:

- Use `WAVE`. Using WAVE, you can view the signals and annotations interactively, and you can correct QRS detection errors (missed beats, false detections, and misplaced annotations) manually if you wish to do so.
- Use `pschart` or `psfd`. Both produce PostScript output.

All four of the detectors above mark all detected beats as normal (N). If your record includes abnormal beats, manually change the N annotations for these beats to the correct annotations. This can be done manually using WAVE.

The information generated by any of these beat detectors can already be imported directly into RHRV using the functions for loading data into WFDB format. However, the for sake of completeness, in this appendix, we will show how to extract information about the beat position and how to interpolate it using the WFDB tools.

B.5 Extracting RR Intervals from an Annotation File

The beat position information generated by the WFDB detectors can be loaded directly into RHRV. However, you can also obtain the tachogram signal, $d_{IT}(k)$, using the `ann2rr` command:

```
ann2rr -r ecg01 -a qrs
```

This command writes the intervals to the standard output in text form. To save the output in a file, redirect it as follows:

```
ann2rr -r ecg01 -a qrs >ecg01.rr1.txt
```

Please refer to the manual page of the `ann2rr` command for additional information [7].

If we are interested in the inverse tachogram signal, $d_{IIT}(k)$, we must use the `ihr` command. For example, use a command such as the following:

```
ihr -r ecg01 -a qrs
```

This command writes the reciprocals of the intervals (i.e., the time series of instantaneous heart rates) to the standard output in text form. To save the output in a file, redirect it as follows:

```
ihr -r ecg01 -a qrs >ecg01.ihr.txt
```

Again, we recommend to read the help page for the `ihr` command for additional information.

B.6 Interpolation

Most (but not all) methods for the frequency-domain analysis of time series require that the series be sampled at uniform intervals. The heart rate time series obtained by ihr is sampled at nonuniform intervals in general. If you plan to study heart rate in the frequency domain, you might want to resample the heart rate signal at uniform intervals. RHRV can perform this interpolation using linear interpolation or splines. You may also use the IPFM algorithm implemented by the WFDB function tach. Use tach to obtain a uniformly sampled heart rate time series with a command such as the following:

```
tach -r ecg01 -a qrs -F 4
```

The argument of the -F option indicates the desired sampling frequency for the heart rate signal, in this case four samples per second. This command writes the resampled heart rate time series to the standard output in text form. As for ihr, the output can be saved in a file by redirecting it. Unlike ihr, the output from tach contains only one column by default (the instantaneous heart rates) since the time intervals between samples of the heart rate signal are uniform.

Both ihr and tach offer powerful techniques for outlier rejection, which is almost always needed when studying long heart rate time series. Read their manual pages for details.

References

1. P.L.L. Sörnmo, *Bioelectrical Signal Processing in Cardiac and Neurological Applications* (Elsevier Academic Press, Cambridge, 2005)
2. W.T.J. Pan, A real-time QRS detection algorithm. IEEE Trans. Biomed. Eng. **BME-32**(3), 230–236 (1985). Available: https://doi.org/10.1109/TBME.1985.325532
3. L.S.O. Pahlm, Software QRS detection in ambulatory monitoring – a review. Med. Biol. Eng. Comput. **22**(4), 289–297 (1984). Available: https://doi.org/10.1007/bf02442095
4. D.L. Farber, Multichannel QRS detector, Master's thesis, Master's degree in Electrical Engineering and Computer Sciences. http://dspace.mit.edu/handle/1721.1/10868 (1996). Last accessed 15 Apr 2024
5. G. Moody. The WFDB applications guide. Last accessed 15 April 2024. [Online]. Available: http://www.physionet.org/physiotools/wag/
6. A.L. Goldberger, L.A. Amaral, L. Glass, J.M. Hausdorff, P.C. Ivanov, R.G. Mark, J.E. Mietus, G.B. Moody, C.-K. Peng, H.E. Stanley, PhysioBank, PhysioToolkit, and PhysioNet. Components of a new research resource for complex physiologic signals. Circulation **101**(23), e215–e220 (2000). Available: https://doi.org/10.1161/01.CIR.101.23.e215
7. WFDB software package. Last accessed 15 Apr 2024. [Online]. Available: http://www.physionet.org/physiotools/wfdb.tar.gz

Index

A
AMI, 82, 98, 100, 101
Analysis, 1–3, 7, 21, 27, 29, 138, 143, 144, 147, 150, 153, 155, 156, 181, 182
    fractal, 80, 90, 93, 120, 121, 123–126
    frequency, 21, 29, 32, 45–53, 60, 62–64, 66, 71, 72, 123, 143, 147, 148, 150, 153, 154, 160, 185
    nonlinear, 3, 21, 29, 32, 79, 80, 93–95, 106, 107, 112, 121, 123–126, 143, 147, 148, 150, 153, 154, 172
    time, 21, 29, 32, 37, 40, 42, 147, 148, 150, 153, 154, 160
ANS, 1–3, 5, 7, 9, 10, 45, 79, 179
    enteric, 5
    parasympathetic, 5, 6, 9, 45, 46, 79, 125, 148
    sympathetic, 4, 6, 9, 45, 46, 79, 125
Atrial fibrillation, 143
Autonomic nervous system, *see* ANS
Average mutual information, *see* AMI

D
Data file format, 21, 26, 160
    ASCII, 22, 23, 25, 26, 133
    EDF, 23
    EDF+, 22, 23, 25, 26
    European data format (*see* EDF+)
    Polar, 23
    RR, 23, 26
    Suunto, 23
    WaveForm dataBase (*see* WFDB)
    WFDB, 22, 23, 25, 133, 182–185

Detrended fluctuation analysis, *see* DFA
DFA, 92, 121–124, 148

E
ECG, 2, 7, 8, 10, 21, 22, 27, 103, 107, 109, 131, 147–150, 154, 179, 180, 183, 184
Electrocardiogram, *see* ECG
Entropy, 85
    approximate entropy, 85, 125, 148
    Kolmogorov-Sinai entropy, 85
    sample entropy, 85, 105, 112–114, 125, 150, 153, 155, 156, 173
Episodes, 23, 37, 40, 42, 121, 131–136, 138–140, 142, 143, 149–151, 153–156, 179, 181, 185

F
fBm, 91–93, 125, 126
fGn, 91–93, 125, 126
Fourier transform, *see* FT
Fractal dimension, 82, 106, 148
    correlation dimension, 82–85, 105–107, 109–112, 115, 116, 125, 148, 172, 173
    generalized dimension, 82–84, 106, 107
    information dimension, 83, 84, 106, 107
Fractional Brownian movement, *see* fBm
Fractional Gaussian noise, *see* fGn
Frequency bands, 45, 54, 65, 70, 74, 163
    HF, 5, 39, 40, 45, 46, 49, 50, 53, 63, 65, 70, 74, 139, 140, 144, 150, 155, 156, 180

LF, 5, 6, 45, 46, 49, 50, 53, 63, 65, 70, 74, 125, 139, 140, 148, 150, 155, 156, 167
LF/HF, 46, 66, 74, 142, 148, 150, 155, 156
ULF, 45, 46, 49–51, 53, 59, 60, 63, 65, 66, 69–71, 139, 140, 167
VLF, 45, 46, 49, 50, 53, 56, 63, 65, 70, 71, 74, 139, 140
FT, 3, 7, 47, 49, 63, 64, 72
   DFT, 47, 48, 57–60
   FFT, 47, 48, 52, 57
   STFT, 49, 51, 63–67, 153
   tapering, 47, 57

H
High frequency, see Frequency bands, HF
HRV index, 7, 9–11, 38, 40, 42, 147, 155, 156, 167
HRVi, see HRV index
Hurst exponent, 91–93

I
Intervals, see Episodes
IRRR, 38–40, 42, 43, 167

K
KS-entropy, see Entropy, Kolmogorov-Sinai entropy

L
Low frequency, see Frequency bands, LF
Lyapunov exponent, see Maximal Lyapunov exponent

M
MADRR, 38–40, 43
Maximal lyapunov exponent, 85, 86, 88, 105, 114, 115, 125, 148, 173

P
Periodogram, 47, 48, 53, 54, 57–60, 123
   Lomb-Scargle, 48, 52, 53, 55, 60
   parametric, 48, 52, 55, 56, 60
Phase space, 80–82, 84–86, 105–107, 110, 114, 116
   embedding dimension, 81, 83–86, 102, 103, 107, 109–112, 172
   embedding theorem, 81, 98
   reconstruction, 81, 83, 98, 102, 104
pNN50, 38–40, 42, 43, 150, 155, 156, 181
Poincaré plot, 90, 119, 120, 125, 151, 172
Power spectral density, see PSD
PSD, 47–49, 52–57, 59, 62, 93, 123

R
Recurrence plot, 86, 87, 116, 117
Recurrence quantification analysis, see RQA
Renyi dimension, see Fractal dimension, generalized dimension
Respiratory sinus arrhythmia, see RSA
r-MSSD, 38, 39, 42, 43, 150, 155, 156, 181
RQA, 86, 87, 89, 116, 172, 173
   determinism, 88, 89, 116
   divergence, 88, 89
   entropy, 88, 89, 116
   L_{max}, 88, 89
   L_{mean}, 88, 89, 116
   laminarity, 88, 89, 116
   ratio, 88, 89
   recurrence matrix, 86, 87, 116
   recurrence rate, 87, 89, 116
   trapping time, 89, 116
   t-recurrence rate, 87, 89, 117, 118
   trend, 87, 89, 116
   V_{max}, 89
RR interval, 21, 24, 26, 27, 29, 32, 37–40, 48, 51, 64, 80, 81, 90, 93, 95, 104, 105, 107–110, 112, 115–117, 120, 122, 123, 125, 148, 150, 151, 179, 181–184
RSA, 2, 5, 46, 51

S
SDANN, 38, 39, 42, 43, 167, 181
SDNN, 37–39, 42, 43, 167, 181
SDNNIDX, 38, 39, 42, 167
SDSD, 38, 39, 42
Short-time Fourier transform, see FT, STFT
Software, 148
   VARVI, 26, 131
Spectral index, 93, 123, 124
Spectrogram, 63, 64, 66, 67, 69, 123
State space, see Phase space
Surrogate data, 80, 95–97

Index

T
TINN, 38, 40, 42, 167

U
Ultra low frequency, *see* Frequency bands, ULF

V
Very low frequency, *see* Frequency bands, VLF

W
Wavelet transform, *see* WT
WT, 50, 51, 63, 70, 72

GPSR Compliance
The European Union's (EU) General Product Safety Regulation (GPSR) is a set of rules that requires consumer products to be safe and our obligations to ensure this.

If you have any concerns about our products, you can contact us on

ProductSafety@springernature.com

In case Publisher is established outside the EU, the EU authorized representative is:

Springer Nature Customer Service Center GmbH
Europaplatz 3
69115 Heidelberg, Germany